Invisible Rivals

Invisible Rivals

*How We Evolved to Compete
in a Cooperative World*

JONATHAN R. GOODMAN

Foreword by Robert A. Foley

Yale UNIVERSITY PRESS

New Haven and London

Published with assistance from the Louis Stern Memorial Fund.

Copyright © 2025 by Jonathan R. Goodman.
All rights reserved.
This book may not be reproduced, in whole or in part, including illustrations, in any form (beyond that copying permitted by Sections 107 and 108 of the U.S. Copyright Law and except by reviewers for the public press), without written permission from the publishers.

Yale University Press books may be purchased in quantity for educational, business, or promotional use. For information, please e-mail sales.press@yale.edu (U.S. office) or sales@yaleup.co.uk (U.K. office).

Set in Minion type by Integrated Publishing Solutions.
Printed in the United States of America.

ISBN 978-0-300-27435-6 (hardcover : alk. paper)
Library of Congress Control Number: 2024950451

A catalogue record for this book is available from the British Library.

Authorized Representative in the EU: Easy Access System Europe, Mustamäe tee 50, 10621 Tallinn, Estonia, gpsr.requests@easproject.com

10 9 8 7 6 5 4 3 2 1

Contents

Foreword by Robert A. Foley vii
Preface xiii

ONE. A Hero of Our Time 1
TWO. Smoke 12
THREE. The Evolution of Invisible Rivalry 44
FOUR. Capital 71
FIVE. The Power of Darkness 104
SIX. Electrification 131
SEVEN. We 164

Glossary 183
Notes 185
Bibliography 201
Acknowledgments 217
Index 219

Foreword

Robert A. Foley

In following him, I follow but myself;
Heaven is my judge, not I for love and duty,
But seeming so, for my peculiar end:
For when my outward action doth demonstrate
The native act and figure of my heart
In compliment extern, 'tis not long after
But I will wear my heart upon my sleeve
For daws to peck at: I am not what I am.

—William Shakespeare, Othello, Act 1, Scene 1

On first acquaintance, Iago could not be more supportive, loyal, and helpful to Othello. But as Shakespeare reveals, Iago is hiding behind outward signs of devotion to pursue his own personal ambitions and destroy Othello. And it is not just fictional men. There also are fictional women: Becky Sharp in Thackeray's *Vanity Fair* is a model of niceness while unscrupulously pursuing her own goals. Literature is replete with such characters: two more are Edmund in *King Lear* or Uriah Heep in *David Copperfield*. The list is endless.

And not, of course, just in literature—in the real world, men and women pursue their own selfish goals while hiding under a cloak of cooperation. These are what Jonathan Goodman calls invisible rivals. They are not necessarily the extremes of Iago, but the day-to-day way in which the twin threads of human behavior—selfish competitiveness and altruistic cooperation are intertwined as strongly as a DNA helix.

One can perhaps blame Descartes with his mind-body dichotomy for making popular dualistic thinking, but it probably is the case that humans are compulsive dualists: good and bad; nature and nurture; biology and culture; Ahura Mazda and Angra Mainyu in Zoroastrianism; even the light and dark sides of the Force in Star Wars. The same is true of many philosophical and anthropological assessments of human nature.

On one hand, we have what is routinely short-handed to Thomas Hobbes's view that, in the natural world, human life is "solitary, poor, nasty, brutish and short," with individuals in a "war of all against all." Only the rule of law, or, more precisely, sovereigns, operates to make human life bearable against the grain of human nature.

On the other hand, equally clichéd is Jean-Jacques Rousseau's view that "Man in a state of nature is essentially peaceful; society is the corrupting influence." Left to themselves, preferably by returning to a simple life, humans would revert to "the noble savage."

Philosophy and anthropology have been the battlegrounds for these two views. Some argue, for example, that war and conflict between groups are the result of recent (in evolutionary terms) transitions, such as agriculture or state societies; others see the origin of violent and selfish behavior rooted deep in our ape ancestry. The stakes in this debate are quite high, because our understanding of human nature and how it can be molded is fundamental to the way we approach other individuals and our attitudes—and those of governments—to how society should be organized.

But these polarized views raise two potential paradoxes. On one hand, if we are entirely selfish, how does our obvious tendency to behave cooperatively mesh with that? Or, if we are truly cooperative, how does our selfish behavior penetrate this prosocial tendency? Jonathan Goodman tackles these paradoxes in this book.

There is no doubt that humans have the most extraordinary ability to cooperate—from something as trivial as tolerating the presence of a stranger sitting next to you on a train, to the willingness of citizens to give money to help others (either through donations or taxation), to the intricate interdependence and trust of combat soldiers, to the ultimate self-sacrifice of those who give up their lives for others. Part of this ability could be explained, as it is for animals, by being related, sharing genes, and therefore evolutionary interests. But the extraordinary thing about human cooperation is that it is seldom among related individuals. Often, but not always, it is people who share a culture, a language, or a social world, but it can even go beyond that; the Live Aid concert in 1985 to support victims of the Ethiopian famine is an inspiring example.

But we cannot simply subsume all our goals to the common good. People do have goals as individuals, and indeed societies generally support these when they do not clash too much with others. If I want to hold the world record for baking the largest cake, this will not impede others, but if I want to park in a disabled space, it will. Or deforest the Amazon, for that matter.

Invisible Rivals is an in-depth exploration of these issues, providing a highly readable guide to the complexity of human nature. Goodman takes the reader on a journey from the way in which deception can run through the fabric of a cooperative society, to how signaling (and receiving signals) is the major axis for evolution, to how cultures arm themselves with morals and codes to protect themselves (more or less successfully) against cheaters, to how cheating can thrive, and to how the cultural and economic context (for example, rising inequality) shapes the tug of war between deceivers and detectors. To support his argument, he draws in a wealth of evidence from animal studies, psychological experiments, and anthropological observations.

Invisible Rivals raises many questions from evolutionary, philosophical, behavioral, and political perspectives. How did the two systems of competition and cooperation evolve? Did visible rivals—the natural world of competition—sink slowly into the shadows as the cooperative behaviors on which human success largely depends were selected for? Or emerge again as a free-rider response to the evolution of prosocial behaviors? Or has there possibly been a long-running co-evolutionary race between cooperators and invisible rivals?

Invisible Rivals touches upon major philosophical concerns. Immanuel Kant thought that motives and intentions were more important than outcomes, a view developed more recently by Elizabeth Anscombe. In contrast, Jeremy Bentham and John Stuart Mill emphasized outcomes rather than motives as the essence of good behavior. Jonathan Goodman's invisible rivals may be expected to flourish in the latter world; if a billionaire donates enough money to eradicate malaria, then his (usually his) reasons—virtue signaling behind a ruthless employment policy—are irrelevant.

The behavioral implications of invisible rivals are among the most interesting and explored extensively. There are effectively—at the broadest scale—three strategies for humans in a world where cooperation is valued. The first is to cooperate—simply to go by the rules and reap the benefits as an individual (reputation) and as a group. The second is to behave selfishly, go against the grain of society, and reap whatever benefits accrue, despite any reputational costs. The third is to appear to be cooperative while actually pursuing selfish goals—to be an invisible rival, in Goodman's terms. This may or may not pay off, but much of what is of interest in the book is how this must generate both ways of disguising behavior and—on the part of others—ways of detecting such a cheater.

Last, there are the practical implications, to which Goodman devotes the end of the book. Humans evolved in small-scale societies, where repeated interactions and face-to-face relationships were the norm. The costs of cheating and the strategies for policing were probably respectively high and easy. The large-scale societies in which most of us live, however, can undermine both of these, eroding trust and opening up negative opportunities. How we find ways to deal with this, particularly in a world where digital invisible rivals might emerge, is one of the major challenges facing people today.

How successful we can be at facing these challenges is not a trivial issue. Explorations, not of the dualistic worlds of Hobbes and Rousseau, but of the subtle interweaving of strategies of cooperation and selfishness that Jonathan Goodman explores so well here, will be central to survival. Othello would have done well to read *Invisible Rivals* to see the real Iago, whose goals would bring him to tragedy.

Preface

Some years ago, I went to a seminar hosted by a biology department in New York. The topic was altruism: How and when do people help others without ulterior motives? Many, if not most, of the attendees felt strongly that the answers are that they don't and never would. We are biologically programmed to be selfish—to survive, reproduce successfully, and focus on what is good for us. Anything else is deception or self-deception.

This reductionist thinking stems from work in biology over the 150 or so years following the publication of Darwin's *The Origin of Species*. The world of biology is full of exploitation, competition, and deception. It makes sense, then, that the world of people, who are biological structures, display those qualities. In the 1970s, this was a popular view in evolutionary theory, and my experiences suggest it is still popular in some groups today.

There is, however, now another movement: one that focuses not on the cold logic of differential reproduction inherent to biology, but on culture, and how culture helped people to break free of the chains of biological evolution. Over many thousands of years, cultural groups—populations of people who share culture—evolved to optimize mutual aid. Those groups, the thinking goes, eliminated or absorbed other groups of selfish people. Cooperation spread because successful culture spread—and here we are, a cooperative species that has conquered biology.

It's hard to attend talks or seminars about either view without becoming depressed. The source of the depression is not about this modern translation of ancient debates around human nature, but in how these remarkable generalizations—or even caricatures—of what it means to be a human are misleading. Coloring a species as diverse and complex as *Homo sapiens* as good or bad—and to me, that's what these debates are really about—is inappropriate, because these models of human nature drive us to create interventions against caricatures. If we decide people are good or bad, competitive or cooperative, selfish or altruistic, or one of any other binary set of variables, we will create policies and enact rules that reflect those decisions. We will create social norms that exist to support or thwart phantoms.

My aim in this book is to convince you not only about the futility of these debates, but also to show how a combination of past theories that presents a more realistic image of what sort of animal a human is can give us the tools to better the cultures in which we live.

This has greater implications than the theoretical. The developing world is likely to feel the modern growth in use of fossil fuels most, which will undoubtedly lead to further extreme weather events in the Global South, as well as countless heat-related deaths. Inequality is rising within rich countries, and evidence suggests that the wealthiest 1 percent of people in the world are responsible for twice the carbon emissions as the poorest half of all people. The burden of noninfectious diseases such as cancer is growing in the developed world; communicable diseases continue to threaten people in less developed countries. There is also a growing epidemic of mental health problems, driven in part by increased loneliness across many cultures.

How we create solutions to these problems must link directly to how we conceptualize human nature. Believing people live in cooperative groups leads some to think we can blame tribalism for everything; believing everyone will always behave selfishly might lead to resignation. Neither will do: what we need is a realistic set of expectations that can inform how we design policy, and (perhaps more important) inform us how we can improve ourselves individually. Understanding ourselves is the first step to solving the problems that are rapidly becoming existential threats.

All I ask is that you read this book with an open mind and question your own biases that drive your reactions. We are likely to have a shared interest in solving the problems enumerated above. How we do so is immaterial. But to take the first step forward, we have to be honest, both with and about ourselves.

Invisible Rivals

1
A Hero of Our Time

In early 1975, two well-known biologists attended the funeral of a colleague who had killed himself. The only other attendees were locals from his community and a contingent of homeless people he had helped in the latter years of his life. He had died after severing his arteries following a bout of depression that flowed largely from the consequences of his work.

His name was George Price. And although he didn't live to see it, his work has had a huge effect on thinking across the world of sciences today. Along with one of the colleagues who'd attended his funeral—a biologist named William Hamilton—Price had developed a theory, encapsulated in a mathematical equation, that helped to explain why organisms help each other. He'd formally modeled the possibility of altruism.

Yet unlike many of his colleagues, Price applied his thinking almost pathologically, possibly because, over his career, he'd become intensely religious. He began to give away all his belongings and to invite people into his home. He then became homeless himself—which didn't stop him from publishing a major paper with the other colleague who later attended his funeral, the biologist John Maynard Smith. Much as George

Orwell sometimes did, during this period Price moved from place to place with homeless friends he'd made; later, he committed suicide.

Price's work and self-sacrifice make him an icon in the history of science. And although we have much to learn from his work, his generosity of spirit, exemplified in his indifference to his own quality of life, shows us one of the extremes of human nature. Many people are familiar with another extreme, selfishness—and yet both tell us a lot about the history of our species, as well as, perhaps, our future.

Over the past few years, I have taught groups of undergraduates about the evolution of cooperation. There are, you may not be surprised to read, various opinions on this subject, reflecting a long history of controversy that predates the application of evolutionary biology to human behavior. One common theme in biology—or more specifically, the subfield dubbed sociobiology in the 1970s by the late E. O. Wilson—is that social behaviors reduce to biological selfishness, or the maximization of reproductive success. Yet over the last two decades, a popular view has developed: we survived, multiplied, and conquered the world because we cooperate more effectively than does any other species.[1]

Debate between these two camps—ones that, broadly speaking, focus on competition and those that emphasize cooperation—are ongoing. What's been more interesting has been how students with whom I have interacted have perceived these various opinions. When I explain some of the theories that rely on people being fundamentally pro- (rather than anti-) social, the most popular sentiment seems to be skepticism.

To me, that skepticism seems justified. Millennials and our younger counterparts, Zoomers, have had an easier time than many before us. We haven't fought in wars or grown up in a society that rations food. Most of our homes are heated. Advanced technology is readily available—and so on. So it isn't that, at least for now, most young people living in democracies such as the United States, Germany, and Taiwan need to fear for their lives every day. The skepticism, instead, is a product of what I can describe only as a deep perniciousness and mistrust that pervade our societies.

Wealth inequality is more extreme than at any other point in the past century. Healthcare is crumbling in the United Kingdom and unaffordable for huge portions of the U.S. population. Extremist political parties across Europe are in positions of strength. In the past few years, the threat of a global war, widespread infectious disease, and populism—on both sides of the spectrum—have returned, after decades of absence. And at a local level, research suggests that narcissists and psychopaths disproportionately hold positions of power, whether as CEOs of major corporations or in politics.

If you watch these events on the news, or on whatever media tool you use, I think you would be forgiven for rejecting academic arguments that humans are a fundamentally cooperative species. We cooperate in a way that keeps society functioning—people typically pay for stuff they buy at the store, and so on—but on many levels, people in positions of power and wealth use what they have for selfish purposes.

And like these students with whom I've worked, I can't help but roll my eyes when people in positions of power write about how good people are when so many social issues seem so bad. Selfishness and prosociality don't define us, but are intrinsic elements of what makes a person a person. We evolved with both good parts and bad, and ignoring either is likely to lead to a lot of problems, at both the individual and societal levels.

Evolutionary theory and its many cousins can help us to explore how. Take, for example, the story of Gravity, a well-known payments company. In August 2022, another Price, this one named Dan, abruptly resigned from his position as CEO after allegations of sexual assault surfaced. That someone in a high-powered corporate position might be accused of assault wasn't new, yet with Price the story made headlines and provoked upheaval on social media. This is because Gravity wasn't any ordinary West Coast start-up—Price had made headlines before. His commitment to fairness had made him famous: he'd pledged to reduce his own million-dollar salary and pay each of his employees at least $70,000 a year.[2]

Positioning himself as an outspoken advocate for ethical business practices was at the core of Price's personal brand. On LinkedIn he'd frequently post motivational content—stuff like "My secret? Great employees!"—that attracted hundreds of thousands of interactions. Coupled with his prophet-like appearance, he'd developed a convincing online persona that led to widespread adulation.

But then the *Guardian* reported that around the Gravity offices, signs started popping up asking, "Have you been abused by Dan Price? We hear you. We believe you. We support you." Soon after, he stepped down when legal action was taken against him for assaulting a woman in his Tesla.[3]

Price isn't, of course, a singular, anomalous example of a person creating the facade of morality to hide whatever they're up to—or use their ethical behavior in one sphere to justify treating others badly. All over the world, from the smallest and simplest amoebas to the most complex mammals, we find similar examples of deception and exploitation. And in a dark way, evolutionary biology is the story of how organisms, humans included, misrepresent themselves for their own benefit.

But the story goes deeper, and different sides about the fundamental morality, or lack thereof, of human nature have appealed to different sensibilities at different points in history.

And a related question is why humans spread across the world—a success that no other species has achieved.

Ever since Charles Darwin wrote his second major work, *The Descent of Man*, people have wondered: what makes humanity so successful? We aren't particularly fast or strong. We don't have natural weapons like claws or poisonous bites. Instead, evolutionary thinkers from Darwin onward argued that our ability to cooperate—part of what Darwin called the "social instinct"—explains our success. We work with each other, knowing that great tasks can't be done alone.

Today, we know that cooperation isn't important just for humans. It is a foundational component for survival in many forms of life. Even tiny organisms such as viruses trade genes to benefit themselves and one another. Bacteria, in a process called quorum sensing, signal information to help individuals adopt phenotypes more beneficial to all members of a given group. And in humans, much research over the past few decades has aimed at showing that leaving a cooperative partner can damage that person's evolutionary success.

Yet competition and exploitation are also universal in the natural world. Cancer cells manipulate the immune system to disguise themselves while thriving off their host's nutrients to survive and multiply. Cuckoos lay their eggs in other birds' nests, forcing or deceiving others into rearing their offspring. And some chimpanzees work to displace or kill dominants in their own social groups, while others falsify alarm calls to gain access to food or mates.

All species exhibit their own versions of manipulative behaviors—and humans, as Dan Price helpfully illustrates, are no exception. Zoological and anthropological research reveals how more complex social groups allow individuals to cooperate better—but also present more opportunities for exploitation.

Related intuitions probably drive skepticism from some

of the people I've taught—and my own skepticism about some popular contemporary views. And what's more interesting: if you look back over the past fifty or so years, you'll find that what people say about human nature seems to track political trends.

For example, older empirical work is consistent with standard social evolutionary theory that developed about half a century ago, where people talk a lot about the idea of *Homo economicus*—the "selfish man" who ignores the well-being of everyone else to focus only on maximizing economic gain. Yet when the giants of modern biology wrote their most influential treatises, selfishness was the prevailing view for explaining all human—or really, all living—behavior in the world of evolutionary theory. Some people even claim that this pessimistic view flowed from trends in contemporary culture: selfishness, particularly economic selfishness, was revered in many parts of Western society, with writers such as Ayn Rand championing self-interest above any other quality.

By contrast, research over the past twenty years in the social sciences has focused on disproving this dark—and probably a bit silly—view of humans. Instead, many people act against their own economic interests for the sake of helping others. And it's because of these findings that many now believe cooperation is our fundamental, immutable characteristic.

This modern, more optimistic view of human nature, which is broadly accepted—if not presupposed—encourages us to get over *economicus* and instead embrace *Homo reciprocans:* the reciprocal person who cooperates with cooperators.[4] For the past several decades, many researchers have advocated for the view that humans are the most, if not the only, altruistic species in the known world. We are, according to such notable biologists as Martin Nowak, "supercooperators," and some supporters of this view seem either ignorant of, or willfully blind to, the ways that exploiters continue to damage society.

Whereas *reciprocans* is an enticing answer to *economicus*, it's difficult not to see in both views the same shortcoming. People aren't just selfish cheaters who expect nothing from others. But that shouldn't imply that self-interested motives don't drive a lot of our behaviors.

Balancing selfishness, exemplified perhaps in Dan Price, and altruism, as shown by George Price, remains a difficult problem for those who want to paint us as fundamentally one or the other. Proponents of *reciprocans* seem to ignore the more sinister side to our nature: that we exploit cooperative systems for our own gain. The broad categories into which contemporary researchers place human beings seem, in other words, to largely ignore our propensity to behave differently in different circumstances.

Invisible Rivalry

We need, instead, a new synthesis of these conflicting evolutionary stories: not *reciprocans* or *economicus*, but a bit of both. We compete and cooperate, and sometimes we pretend to cooperate only to compete more effectively. This is underpinned by a broader, more universal set of evolved behaviors that encompasses not just the struggle for survival, mates, and wealth, but the struggle for social success and for power over others. Because of our extreme social intelligence and ability to use language effectively, we can hide the strategies we use to promote our own interests without anyone, perhaps even ourselves, realizing our goal. We strategize our way to the top of our social hierarchies. We are, to others and sometimes to ourselves, invisible rivals.

The blur between the old, shallow models of humans as bad or good, selfish or altruistic, competitive or cooperative, creates a vacuum across the psychological and evolutionary sciences with serious consequences for society. Ultimately, the thesis that humans—like all animals—are fundamentally selfish follows from Darwin's earliest theories of evolution by natural and sexual selection. The antithesis, that we overcame that selfishness through mechanisms like punishment, and consequently domesticated our wild minds, is many researchers' cure for that ancient disease.

Neither is enough. Self-interest maximization, an instinct with which all organisms are born, merely evolved, much like a cancer, taking on a new, perfidious form—invisible rivalry. We didn't rid ourselves of selfishness, we just got better at hiding it.

Just as we've spent the past century developing novel ways of destroying cancer, we've spent our entire history trying to rid our social groups of free riders, defined as people who take from others without giving anything back.[5] Cancer does many things to thwart oncologists, the professionals who study and treat it. It hides, it changes, and it uses our own weapons against us. How the analogous problem manifests in human groups depends on the relative culture and norms, but the rule that if a system can be exploited, it will be, holds across organisms and societies alike.

Even science, which in theory is a dispassionate subject, is not immune to exploitation. Many people, for example, point out a major consequence of how communist ideology infected scientific reasoning during Stalin's regime. Trofim Lysenko, the infamous agronomist who rejected the tenets of modern genetics, invoked the largely debunked theories of Jean-Baptiste Lamarck, an eighteenth-century French naturalist, to defend his views. What followed was the advent of Lysenkoism, which basically relied on Marxist political thought in informing how evolutionary change takes place. His thinking was not only wrong, but fatal: the use of Lysenkoism in crop production led directly to widespread famine in both Soviet Russia and China.

Interestingly, Vasily Grossman, the twentieth-century journalist and novelist, wrote an analogous story in his major fictional work, *Life and Fate*. The way Grossman depicts the life of a scientist in Stalinist Russia, however, makes the simple vilification of someone like Lysenko more complicated. Amid threats of harm to their families, worries over denunciation, and other common punishments in the Soviet Union, many scientists in Grossman's story change their views to avoid harm. This does not justify Lysenkoism, but it does show, convincingly, how even some good scientists can bend their will for the state.

Still, people today comfortably refer to Lysenko as a pseudoscientist, and it is easy, with a view from modern, Western culture, to see a direct line between Lysenko's political beliefs and starvation. Yet for anyone, it is difficult to try to remove yourself—scientist or not—from your own biases, and see to what degree you are being influenced by external forces in analogous ways. We are all victims of political trends, and it is no coincidence that the way academics interpret human nature appears to track what is happening in society more broadly. When Milton Friedman, the economist and Nobel laureate, wrote his major defense of free-market capitalism, for example, major biologists in the same decade also argued for the fundamental selfishness of all organisms. And as culture—particularly progressive culture—has leaned toward the pursuit of equality and sustainability in economic thinking, it is notable that many academics now argue that despite our humble beginnings, human culture has allowed us to overcome selfishness in favor of cooperative social norms. An academic article in 2023 suggested, to this point, that "soft censorship"—or concerns that publishing against popular views will harm your academic career—harms dissent in the world of research.[6]

More generally, in Western democracies, we don't (for now) have tyrants forcing their political philosophies into our daily thinking. But we do have broader social movements that determine how we present ourselves to others—and not always for the best. When Justine Sacco, a public relations executive, sent a widely shared and vilified tweet in 2013 saying "Going to Africa. Hope I don't get AIDS. Just kidding. I'm white!," thousands of people cooperated to create the world's first cancellation on social media.[7]

Sacco ended up losing both her reputation and her job. And yet, as the journalist Jon Ronson pointed out, no one punished—to use the language of anthropology—another person who tweeted they were "actually kind of hoping Justine Sacco gets AIDS? lol."[8]

Social media are extreme examples of platforms that tend to drive passions to their highest, and some anonymity gives people the power to be hateful when, in person, they might just complain about the weather. That is not a new point. The broader idea, however, is that, in digital and analog communication alike, political and social trends drive belief and action. We are not free of the same forces that led to Lysenkoism. Instead, the way we are influenced to adopt beliefs, academic or otherwise, reflects the peculiarities of our cultures—and as cancel culture shows, the need to signal cooperation with others about a given trend can lead to virulent hatred.

Yet we can try to view ourselves objectively, much as we view Lysenkoism from an outsider's point of view. We're capable of the most extreme forms of cooperation in the known world—but that's an ability we should laud, not presuppose. To paraphrase Karl Marx, the point isn't just to understand human nature, but to change it. And adopting a synthetic understanding of human competition will help us to adjust our expectations from people, and create rules that bridge self-interest with the interests of others. Evolution endowed us with the intelligence to overcome selfishness—and we should use it.

If you are skeptical about the idea that humans aren't fundamentally cooperative, please read some of the evidence I offer before deciding against me. If you think people are all just rotten, the same plea applies. And if you just aren't interested in reading a book about science, relax with the knowledge that this is not a work of science, but a discussion about people that uses some input from science, and much other input from elsewhere. The question of how to apply a synthetic understanding of human nature—relying on anthropology, sociology, and current events—to help fix issues like poor public trust and inequality can't be answered by one discipline, or even a few. It lies in understanding ourselves and one another, a task that science cannot do alone.

Just as Jonathan Swift wrote that humans are not rational animals, but rather "something altogether more tentative and difficult, an animal which has the capability of reason," we are similarly not cooperative animals, but animals capable of cooperation.[9] Recognizing this distinction allows us to promote working together, rather than to presume it—and to build our societies with our flaws in mind.

2
Smoke

All smoke and steam . . . all seems forever changing, on all sides new forms, phantoms flying after phantoms, while in reality it is all the same and the same again; everything hurrying, flying towards something, and everything vanishing without a trace, attaining to nothing.

—*Ivan Turgenev,* Smoke

People who develop cultlike followings tend to have a few qualities in common. They are usually captivating—sometimes literally—intelligent, and apparently devoted people who advertise effectively their determination to improve the lives of those around them.

Yet they also, intuitively or otherwise, understand what vulnerabilities people have, and use those vulnerabilities against them—often without their victims realizing it. Robert Hendy-Freegard, for example, a British con artist, was working at a pub in a rural county when he told several students he was an undercover MI5 operative. He tricked three people—one man and two women—first into working for him, and next into fearing that their undercover identities had been revealed and that they needed to run away. He alienated them from their friends, persuaded them to give him money, and eventually abandoned them. Hendy-Freegard, it was later reported, seduced an unknown number of women over his career of conning. So effective were his deceptions that British tabloids started calling him the "Puppet Master."[1]

Cult leaders in the more colloquial sense use analogous strategies, though channeled differently. In 1973, Robert G. Millar founded Elohim City in Oklahoma. Millar was an advocate for, and leader of, the Christian Identity movement, which claims that only those with Aryan heritage are the chosen people of the Old Testament, and consequently that these people, all of European ancestry, are the living descendants of the ancient Israelites. (There is, unsurprisingly, a deep thread of anti-Semitism in the movement.)

For decades, Millar acted as leader of Elohim City, which was connected to right-wing extremism in the United States; polygamy was also permissible in the community. Today, about one hundred people live in what is essentially an armed compound—a living testament, as it were, to the religious tenets Millar espoused.

These are two apparently unrelated examples of how one person—someone who appears devoted to some movement—can attract several if not dozens of followers to a cult, loosely defined, of their own personality. The psychological tactics are interesting but not relevant here. Instead, the salient parallel is in both Hendy-Freegard's and Millar's sexual and reproductive success. Hendy-Freegard mostly convinced women that he was an undercover agent, and conned them into giving him money and running away with him—and occasionally, even having children with him. Millar, according to the Southern Poverty

Law Center, is father to eight residents of Elohim City, and grandfather to fifty-eight.[2] (If one hundred people live in the community, then Millar is a direct genetic progenitor of 64 percent of the population.)

These details are not coincidences. People with charisma who lead others away from their former lives tend to have a lot of sexual or reproductive partners. Fred Newman, a former philosophy lecturer in New York, founded the Newman Tendency, a cult that promoted what he called friendosexuality—sexual relationships only with in-group members—and Newman himself was reported to be having affairs with eighteen cult members at a time. David Koresh, who led the Branch Davidians in a movement that culminated in the tragic standoff with the FBI in Waco, Texas, justified taking dozens of "spiritual wives" through a biblical prophecy. And Shlomo Helbrans, a former rabbi who formed a Jewish cult that he led, eventually, to South America to dodge authorities after allegations of child abuse, has had multiple children with members, one of which now rules the group, known as Lev Tahor ("pure heart" in Hebrew), using horrific tactics that include torture, starvation, and murder.[3]

Dr. Alexandra Stein, a social psychologist who herself escaped from a shadowy cult known as The Organization (or "the O" for short), wrote in 2017 that the point of these groups isn't always sexual exploitation—although that is a common theme—but rather total control of the adherents' beliefs and actions.[4] While in the O, Stein had a contact who gave her notes explaining what she could wear and whom she could marry. Members of the Newman Tendency frequently needed to prove themselves to other members through various tests. And those unlucky enough to be born into the Lev Tahor were forced to memorize the works of Helbrans—at the expense of any other form of education—or else to, among other privations, swallow hot powder, and thank their torturer for hurting them.[5]

From an evolutionary point of view, what's interesting, if creepy, about these groups and their leaders is how followers are torn away from their families and social connections. The ostensible secrets the groups offer to solve attract people who feel lost in society. And for these secrets, these followers often sacrifice the protections—however limited—that society otherwise offers them.

In a fundamental if incomplete way, social norms exist to help reduce the risk of people being exploited by others. Cults and free-agent con artists like Hendy-Freegard are effective at pulling people away from those norms for their own ends, whatever those might be. And they are undoubtedly a product of large-scale societies where it's easy to feel alone or alienated, or that you're missing some big secret other people are in on.

These forms of deception lead to a critical question: do they represent something new and weird about the modern world?—or do their roots go further back?

To attempt an answer, we need to look back at the roots of our own societies. *Homo sapiens* evolved about 300,000 years ago, which is the current best guess. Many qualities distinguished us from other primates—our diets, our behaviors, and of course, our big brains. People have been arguing about which of these drove our social conquest of this planet, though it's unlikely that we'll ever be able to point to a specific quality of humans and declare "this is what made us special."

Nonetheless, our unique disposition to create elaborate cultures over time stands out as an important feature of even ancient human groups. Stone tools—artifacts that are grouped under the heading of material culture in archaeology—millions of years old have been discovered, and were used by our ancestors before *Homo sapiens* evolved. Research into both ancient and modern humans shows the breadth of cultural diversity a single species can create in what is, at least in evolutionary terms, a short time. An analysis of this rich cultural history can, then, help us answer the key question: did ancient societies rid human groups of deceptive exploiters—or is any story of this kind just smoke, behind which our true nature lies?

In Search of the Noble Savage

One feature of humans people repeatedly point to that purportedly explains our success is cooperation, which in biology roughly translates to making an effort to help someone else.[6] We cooperate with others—crucially, nonrelatives—to a greater degree than any other known species. And many people think that our predilection for cooperation is the core of human nature.

The French philosopher and essayist Michel de Montaigne, for example, argued in 1580 that ostensibly advanced societies, like many of those at the time in Europe, were no less barbarous—a word that was popularly used for basically anything not European—than those of the newly discovered New World. His account is taken from an "ignorant" man who spent some time in the Americas. And, Montaigne adds, he believed that man, saying that plain people are more likely to tell the truth. Of the "better-bred types," who so contrast themselves with plain folk, he says: "they never represent things to you simply as they are, but rather as they appeared to them, or as they would have them appear to you, and to gain the reputation of men of judgement."[7] Highly political cultures such as those in sixteenth-century Europe, for Montaigne, led to distortions of truth for personal gain. Societies without these pressures, by contrast, did not practice deception—or even understand the concept.

His assertions about the then newly discovered cultures of the Americas are of course preposterous. But nonetheless he saw in his own world such flaws that he felt weren't recognized by his contemporaries, and moreover that Europeans were dogmatic about the superiority of their cultures, despite the deceptive machinations that lay beneath the veneer of sophistication.

This insight led the Swiss philosopher Jean-Jacques Rousseau to popularize Montaigne's thinking in the form of "the noble savage."[8] This view, which probably every political science and philosophy undergraduate is tortured with at some point in their formative years, has been foundational to thinking about preindustrial human society for the past several centuries. Rousseau's view is that, devoid of the insidious facets intrinsic to large, technologically sophisticated societies, people are fundamentally generous and kind to one another. It's the state that messes up things—the rules and intrigue that pervade larger groups, where we can't know everyone else personally, subvert the natural kindness intrinsic to human nature.

Rousseau's writing led to the popular view of libertarianism—the idea that without a state subjecting people to rules and norms that result in deception, intrigue, and exploitation, the natural, anarchic form of human living takes hold. And in this state without a state, people are left with the freedom to be kind. Modern libertarianism seen in some Western democracies is a descendant of this philosophy of human nature. It's also an attractive view: to think of our ancestors as lacking the traits we find so dislikeable today, we can aim for a better world by pushing for a smaller government.

Anarchists of the nineteenth and early twentieth centuries

embodied this desire. Karl Marx and Friedrich Engels are two major thinkers sometimes counted in their ranks—but Pyotr Kropotkin, a Russian aristocrat and explorer, is perhaps the least well-known, and indirectly most influential, anarchist thinker of the time. Kropotkin, who was born in the mid-nineteenth century, became interested in Darwin's ideas early in his career. In the 1860s, he, along with his friend the zoologist Ivan Poliakov, trekked across eastern Siberia—one of the coldest, least hospitable climates in the world—to learn about how best to interpret Darwin's theory of evolution by natural selection, which had first been published less than a decade earlier.

They were determined—against the scientific hegemony emanating from the United Kingdom at the time—to prove that natural selection was not, to borrow a phrase of Tennyson's, red in tooth and claw. And because of these interests (or, more strongly, biases), Kropotkin found what he was looking for. In his major work, *Mutual Aid: A Factor of Evolution*, from 1902, he wrote, "We both were under the fresh impression of the Origin of Species, but we vainly looked for the keen competition between animals of the same species which the reading of Darwin's work had prepared us to expect . . . but even in the Amur and Usuri regions, where animal life swarms in abundance, facts of real competition and struggle between higher animals of the same species came very seldom under my notice, though I eagerly searched for them."[9]

Poliakov and Kropotkin felt, instead, that the struggle they observed was all against common enemies: climate pressures, predation, and so forth—all of which are (famously) intense in Siberia. This idea, relating to Darwin's of "hostile forces," ties into another view wrongly attributed to Darwin: that individuals act naturally for the good of their respective species. Karl Kessler, the nineteenth-century Russian zoologist who heavily influenced Kropotkin, declared in 1880 that "All organic beings have two essential needs: that of nutrition, and that of propagating the species. The former brings them to a struggle and to mutual extermination, while the needs of maintaining the species bring them to approach one another and to support one another."[10]

This statement contrasts with not only the thinking of Darwin and his followers, but also basically all consequent major advances in biology over the subsequent century. Yet from the unification of Darwin's theories with Gregor Mendel's newly discovered science of genetics, all the way through to the 1970s and even beyond, groups of scientists formed factions around this question: do organisms behave to maximize their own biological success or for the success of their species?

Despite some current protestations to the contrary, the theory of evolution by natural selection focuses on whether an individual within a species successfully reproduces. Biologically speaking, individuals act for their genetic lineages, not for the greater good of their species—though of course whether a species survives depends on the survival of the individuals that compose it.

It's notable that Kropotkin highlights the unfriendliness of the Russian geography and climate, which probably forces greater cooperation among successful organisms, in support of his observations.[11] And perhaps it was these factors, combined with evolving international politics—the first volume of Karl Marx's *Das Kapital* was published in 1867—that led Kessler, Kropotkin, and Poliakov to their conclusions.

Unlike Marx, however, Kropotkin used his insights about human and animal nature to justify a defense of what became known as anarcho-communism: he believed that ruling bodies repressed the mutual kindness inherent to all uncorrupted organisms. This was distinct from other views becoming popular across Europe, and Kropotkin was particularly influenced by an anarchist group of watchmakers he met when he visited the Jura Mountains of Switzerland in 1872.[12] There, he became impressed by the local anarchists' steadfastness to the international movement: they funded, for example, union strikers in the United States through their wide-reaching network. They also banded together through mutual support to push back against the oppressive working conditions in the watch-making factories that employed them—and even against Marxist factions that many felt, even then, might lead to authoritarian states. Later in his life, he viewed the Bolshevik regime as just another form of state corruption, and remained critical of all governments, including the authoritarian variety of his home, until his death in 1921.

Although Pyotr Kropotkin isn't a household name like Karl Marx, his ideology has had far-reaching consequences for both anthropologists and non-academic political activists, including modern libertarians and socialists. But his notion of mutual aid rests upon a critical premise: that in their natural state—whatever that is—animals help each other in common struggles. This notion of mutual aid leads to a question that has been dressed and undressed in different cultures and languages too many times to quantify: how would humans behave toward one another if we were to strip away the oppressive rules common to all large societies?

Ethnographic research—and anthropology more broadly—aims, in part, to answer this question. In learning about the origins of human society, we can understand whether there are universals to what makes a human a human. And yet despite this noble aim, these social studies have been infected with ideology since their creation, biasing our interpretations of data in the science of human nature. We are all guilty of importing our own forms of Lysenkoism to our respective worlds.

This is typified in much anthropological thinking over

the past century. The noble savage is everywhere in studies of ancestral humans, dressed up in the language of egalitarianism. For example, a popular view is that human societies experienced a U-shaped relationship with equality. We were, before the development of complex culture, hierarchical in our groups: large, aggressive alphas dominated subordinates, had many mates and offspring, and controlled territories with violence.

Yet some prominent academics—notably Richard Wrangham and Brian Hare—argue that, just as people domesticated dogs to be our companions, so did we domesticate ourselves over the past ten to twenty millennia by ridding ourselves of the overtly violent and domineering people who used to dominate our social groups. Wrangham calls this selection against reactive aggression—the subtype of aggression that, despite how it sounds, was found most often in the alphas that headed groups of our ancestors.[13]

Archaeological research supports this view. As hominins evolved over the last several million years, we began experimenting more and more with the world around us. Chimpanzees, our closest genetic relatives, use tools like sticks for termite-fishing—but our ancestors, even predating the *Homo* genus, were creating stone tools as long ago as 3.3 million years before the present.[14] Artifacts found at the Lomekwi 3 site in modern-day Kenya, which include a fifteen-kilogram anvil, suggest that our ancient relatives were, even then, developing larger, more complex tools than those the intelligent primates who make up the genus *Pan*—such as chimpanzees and bonobos—do today.[15] Stone tools dating from this period through the near present confirm that, across the globe, we've been manipulating the world around us for a huge length of time over our evolutionary history.

By the time *Homo sapiens* evolved, hominins had been mastering stone tool-making for millions of years. And as our

Stone tools found in North America dating to 8000 BCE to 1600 CE (Smithsonian Museum, Anacostia Community Museum Collection, CC0)

cultures evolved, we started levelling the fighting field against aggressive alphas: weapons and poisons helped smaller, more intelligent people kill the dominants that oppressed us.[16] Sometime over the last twenty thousand or so years, according to Wrangham, we toppled these rulers and rid ourselves of reactive aggression—leading to a Golden Age of egalitarianism in human prehistory. (The idea of the U-shape relates to the presence of inequality across much of the world today.)[17] So how can we find these noble savages?

There are two ways researchers have traditionally aimed to learn about what our historical societies looked like. The first is archaeological investigation: finding ancient sites, carefully reconstructing artifacts like stone tools, and deciphering elements of material culture, such as arrowheads and artwork. The second is ethnographic study: forming a picture of what ancient society looked like by immersing ourselves in the hunter-gatherers living today (or consulting records of these immersions by others).

Both come with advantages and drawbacks, but typically, over the past century or so, anthropologists have used data collected from ethnographic studies to justify universal claims about what prehistorical societies looked like. The Aché of eastern Paraguay, in South America, are one hunter-gatherer society many people have cited for their egalitarian social norms. Although men are the main food providers among the Aché, the hunters themselves are largely forbidden from eating the meat they bring back to camps themselves. And whereas they are structurally a monogamous culture, the wives of successful hunters—and their children—don't appear to eat any more meat than others. Sharing, which is mostly based on the need of the recipient, is, instead, the norm that governs much of the caloric intake among these people.[18]

Similarly, the Nuer people of southern Sudan were notably equitable in their internal relationships. Although there is not much difference in prestige among male adults, households have leaders or "bulls"; close relations among these households also mean that most members of a village claim relation to the village head, reducing the degree of nepotism in Nuer social groups. If everyone is related to a leader, few, if anyone, should get special favors.[19]

Across the ethnographic atlas, you're unlikely to find many groups of humans with a domineering male leader, which you would find in groups of, say, gorillas. Instead, you'll find sophisticated social structures with intricate social norms around kinship relations and sharing. Among the Maasai people of Kenya, for example, sharing is frequently between people who share a bond called osotua, which literally translates to "umbilical cord."[20] The bond usually forms when one person—as with the Aché, out of need—makes a request of someone else.

An ethnographic study from the 1960s described how the strength of an osotua bond transcends even death—where one member of the partnership dies, that person's children become the replacement. Interviews with Maasai people illustrated just how much like a physical bond osotua is: many said nothing could end the relationship, and exploitation—that is, asking for something someone doesn't need—is "unthinkable."

It may be that need-based sharing systems work so well because it's so difficult to hide resources in hunter-gatherer societies. You can't hide a cow—or at least, not without difficulty. Yet my colleague Athena Aktipis, who researches the evolution of cooperation, thinks sharing based on need can work just as well in large, industrialized societies as in smaller communities like those of the Maasai: "I think they are actually pretty resilient to exploitation in general because they typically ratchet up slowly and are based on relationships which take time and energy to build and maintain," she wrote to me recently. "And most of us have need-based transfer . . . partners even if we don't call them that. Your kid's godmother, your best man, your childhood best friend. For most of us, those are people we could call up in a time of need and they would be there to help us."

Most, if not all, communities practice what scholars like Aktipis call "risk-pooling." We spread risk across groups so that, if one person has bad luck, the others can make it up. The Rossel Islanders of Melanesia, for example, have lived in a relatively isolated way for as long as thirty thousand years, for geographical reasons. Their language, Yélî Dnye, appears unrelated to other existing languages, and their culture has some features that don't seem to be shared with other groups in the region.[21] This makes the Rossel people—who number as few as thirty-four hundred—particularly interesting ones to study: by learning from people who have had little interaction with the outside world, we can see more clearly into what ancient cultures really may have been like before they became adulterated by encountering technologically advanced societies.

Once or twice every ten years, a powerful cyclone hits Rossel Island, which can have disastrous effects for the islanders' homes and resources. In 1849, the anatomist Thomas Huxley (sometimes called Darwin's bulldog in historical letters) observed what appeared to be cyclone shelters on the island.[22] Anthropologists have since learned that locals believe these shelters were built by their central deity, *Ngwonoch:a,* though they are maintained by the populace. When a cyclone strikes, the locals retreat to them, and thereafter some unique cultural features emerge. Ownership, when these disasters occur, becomes irrelevant, and small groups form within the community to retrieve essential foods from the environment, such as palm starch. Without these cultural innovations—the maintenance of shelters and the quick, group-level responses to disaster—these communities would have vanished in such an inhospitable environment. But the Rossel Islanders aren't alone in their risk-pooling systems.

In hunter-gatherer societies, such systems are critical. Where big game animals, such as antelope, are important for human nutrition, the success rates in hunting are quite a lot lower than you'd expect in some cultures, even for the best hunters. The Aché bring home meat on nearly half of their hunting days, but the !Kung (Ju/hoansi) people of the Kalahari Desert successfully do so only about once every four days. Among the Hadza people of Tanzania, the data are harder to obtain, but estimates vary between once every month and once every three months.[23]

Wolunga Bay, Rossel Island (from W. E. Armstrong, *Rossel Island: An Ethnological Study*, Cambridge: Cambridge University Press, 1928, Plate IIB)

It's reasonable, then, to infer that sharing isn't only common, but necessary. For a society to thrive, unsuccessful hunters need to be supplemented by successful ones—and by other means of provisioning, such as gathering and small-game hunting. We end up, in many examples of hunter-gatherer groups, with wide sharing networks that help to provision less lucky—or less skillful—hunters.

There is an obvious nobility to these systems. Each is specific to the local environment and local populace, and within each, people rely on one another when times are difficult. But what isn't clear is the degree to which sharing is equal.

Anthropologists have spent a lot of time, money, and effort trying to understand this. And although it's clear from multiple cross-cultural surveys that people don't just hoard whatever they can for themselves, they do think carefully about how they share—and more important, with whom they share.

One study, conducted by my colleague Nikhil Chaudhary and others, explored this question explicitly. Many hunter-gatherer groups are fond of honey: it's high in sugar, high in calories, and—according to some people—tastes great. Their love of it, in some places, leads them to risking their lives in a way that I think few people would—for anything. Footage taken from research into the Twa—former hunter-gatherers living in Uganda—documents a man climbing a two-hundred-foot tree to hack away at a beehive to retrieve honey, without ropes, and being stung nonstop during his efforts. His wife describes how many men fall to their deaths in similar attempts. The Twa are not alone in this, and research over the past decade suggests that tree climbing for valuable foods such as honey has been foundational in human evolution.[24]

Honey is, then, an obviously valuable commodity in many hunter-gatherer groups, such as the BaYaka people of Central Africa. Chaudhary conducted his research with the BaYaka with this assumption in mind. He and his colleagues asked camp members to distribute sticks made of honey among their campmates, and found evidence that the sticks were not distributed equitably: people showed preferences in how they made these valuable donations.[25]

Several other findings were notable. First, people didn't seem to distribute more to family members than to others, suggesting that the participants' motivations were not nepotistic. And second, there was some evidence that preferences passed to people's children: someone who liked a campmate's father might be more likely to donate to his child. Chaudhary notes that the overall effect he observed is called "relational wealth," and suggests that having a rich social network to rely on is

Climbing can be extremely dangerous and requires bridging techniques to avoid falling (Bruno Zanzottera/Parallelozero)

critical—perhaps even more important than having resources of your own.

These results mirror those of many others that explored economic relationships—known as economic games—in hunter-gatherer societies. In cross-cultural studies, a key finding is that participants want to know something about where their donation is going. Hunter-gatherers don't want to donate resources to an anonymous source—they want to know about their relation to the recipient. Are they family? Are they a hunting partner? Is my family connected to theirs through marriage? These relational elements govern many, if not all, of the social decisions made among these people.

It seems strange, then, to argue that ancient societies were equitable—though in *Mutual Aid,* Kropotkin gives numerous examples he regards as proof of "primitive" egalitarianism. Yet even though ancient humans might not have lived under a state, corrupt or otherwise, that demanded taxes in exchange for protection or other rights, they also probably didn't share precious resources equally. When there was no food surplus, relational wealth may have determined who survived. Beneficence depends, as the Scottish Enlightenment philosopher David Hume argued, on the availability of a resource. And unequal distribution also suggests that people living without a state don't, as Kropotkin believed, act for the good of the species. They act for the good of those they like.

Yet despite protestations to the contrary by many modern thinkers, the history of ethnographic work doesn't support Kropotkin's views either. Almost half a century ago, an anthropologist named John Moore wrote a pair of essays attacking what he called the libertarian view—the idea that, in an anarchic state, people would treat each other equally.[26]

Against that view, popular both then and now, Moore argued that exploitation is an essential element of all human societies, ancient and modern. Oppression and coercion, two favorite topics for people with libertarian ideals, clouded the inconvenient truth that exploitation occurs at every level in society. People in control of a powerful state aren't the only exploiters.

For evidence, Moore consulted a series of ethnographic studies from the preceding century or so. He found, in his survey, a depressing frequency of men exploiting women in most hunter-gatherer societies across the world—an uncomfortable result for those of us who want to believe that ancient humans lived in a state of equality. About one Native American tribe, for example, Moore wrote: "A Cheyenne man could put his wife 'on the prairie' if she committed adultery or was disobedient, or simply to change his luck in war or hunting. . . . A woman put 'on the prairie' was gang-raped by the members of her husband's soldier society, and sometimes beaten and killed. Women treated in this way were 'bad wives,' women who did not live up to their responsibilities in Cheyenne society."[27]

He went on to argue that women's responsibilities—notably, according to one source, before the arrival of Europeans in America—entailed doing just about everything that isn't violent. They would take down and carry tipis; they would set up tipis in new locations; they would cook, cut down trees, and gather wood. If a man did any of these things, people would laugh at him.

Eleven of the fifteen societies he evaluated showed at least the apparent exploitation of women and evidence of older men exploiting younger men. And even stranger, some of the same sources that Kropotkin used to justify his philosophy of mutual aid depict abusive exploitation by the elderly—a social structure known as a gerontocracy. Among the Khoekhoen people of southwestern Africa, for example, a seventeenth-century source described the chiefs as having the "power of life and death," and reported that the elderly urinated on young men and insulted them by calling them women.[28]

The source, a Dutch explorer named Johannes Gulielmus de Grevenbroek, wrote, "Should any of our Hottentots refuse to subject his male members to the sacrificial blade or lancet or operating knife of the priest, preferring to preserve his genital organs perfect in the shape and number provided by nature rather than submit himself to agonies of pain and partial castration, this enemy of all amputation is insulted... shut out from all fellowship and inheritance, and shunned as if blasted by the lightning of heaven."[29] Women, likewise, were "abused like cattle or slaves," and their legs were kept bound to prevent their escape from their local groups.

Of some Australian aborigine tribes, Moore described gerontocratic structures where men over forty monopolized both the marriage market and the distribution of food. They also forced younger men to undergo circumcision and mutilation, and frequently changed out their older wives for younger ones whenever they wanted.

A key point for Moore is also that, in the sample of societies he reviewed, there were no examples of women exploiting men. In some cases—such as the Sirionó of eastern Bolivia—there was no evidence of men exploiting women, though even there, the anthropologist Allan Holmberg writes that chiefs and good hunters tended to have more marriages, with a greater number of wives being a marker of status.[30] But if egalitarianism—in the intuitive sense that people treat each other as equals—were the resting state of human nature, we'd expect as many cases of women exploiting men as men exploiting women. No such balance exists.

Nonetheless, despite this convincing set of arguments, the orthodox view among many people studying human evolution has been that we were, in our ancestral state, egalitarian, though many writers add that this applies "only with respect to people of the same age and sex."[31] This view is taught widely in anthropology courses in Europe and the United States, with the understanding that small-scale societies are effective at suppressing the overt selfishness we tend to associate with the modern, industrialized world. The arguments of twentieth-century thinkers like Moore have largely been forgotten.

Today, however, the picture is changing again. Young anthropologists are questioning the orthodox view, sometimes called the "nomadic-egalitarian model." That view, according to Manvir Singh and Luke Glowacki, the authors of an academic article in 2022, stems from "Man the Hunter," a mid-twentieth-century symposium that inaugurated the field of hunter-gatherer studies. The symposium's organizers defined foraging societies using five criteria:

- First, nomadic societies are egalitarian, defined either as having fairly distributed resources, such as food, or as freedom from coercive control.
- Second, these societies are small enough that people could know one another individually, maximizing the odds that people would cooperate.
- Third, groups lack distinct territorial boundaries, removing the risk of fighting within groups for land.
- Fourth, people don't store sources like food, so that it wasn't possible to monopolize any resource.
- And fifth, the society has a fluid structure that permits people to come and go, which helps to reduce aggression within groups where social networks break down, or where groups grow too large.[32]

These qualities sound ideal for optimizing the kindness with which people can treat each other. And in some hunter-gatherer groups such as the Aché, we see a huge amount of fairness and beneficence among fellow group members. And yet the picture is more complex, as Singh and Glowacki argue—and which Moore's research also suggests. Instead of being entirely egalitarian or entirely warlike and exploitative, there's evidence of diversity in ancient human societies, and we're unlikely to be able to point to any one ritual or cultural practice and say that all groups of humans have it in common.

It's often the environments in which we live that contribute to the social structures that people form. Some are, under Western eyes, evidence that humans evolved to be kind. And some are not.

Among the Aché, people practice wide food sharing, and it's rare for any person or family to be refused food after a kill. There aren't coercive political structures; people don't discriminate against others. If you were an anthropologist working with the Aché, you'd be forgiven for thinking that foraging society is vastly preferable, in many ways, to how we live today.

Other examples are, like those that Moore cites, less savory. In addition to the well-established examples of gerontocracy and patriarchy among people in Australia, exploitation and coercive control have been noted by anthropologists in populations of foragers living in areas like the Bering Strait and western Siberia. One ethnographic report of the Khanti people of Siberia gives evidence of elders using poorer people "like slaves." And a review of ethnographic surveys suggests that the Khanti are not outliers, though foragers are less likely than other societies to exercise coercive control.

Historical records mention the Calusa—which apparently translates to "fierce people"—sedentary foragers who lived in the area of modern Florida from around the year 800 to 1550. They relied largely on seafood, and there's evidence that they managed resources carefully through a centralized state, which was led by a king, which was a hereditary position. There was evidence of a class system, and a distinction between "commoners" and "nobles," and records suggest systemic inequality. The Calusa had a military that even allowed them to repel Spanish influences for two centuries—longer than any other Native American group. Captives the Calusa took from other groups were, according to one source, used as slaves or for human sacrifice.[33]

In fact, of thirty-four sedentary or semi-sedentary societies Singh and Glowacki evaluated, seventeen showed evidence of inequality, whether in resources, status, or wealth. The five criteria that ostensibly applied to forager societies didn't appear universal—or even in the majority of cases. Some were large, some were small; some stored food, some relied on what they found in the shorter term. The only universal feature of foraging cultures, it seems, is that every culture has its own, ecologically specialized set of foraging practices.

The Mode of Production

One important feature when analyzing human societies, following insights from centuries of ethnographic work, is how people subsist and survive. No two areas are equivalent in the kinds of resources people and animals obtain from them, or in how those resources are obtained.

Most animals that migrate to new places either die off or evolve into a new species. Ants, for example, have spread across the world—and, by numbers, they are among the most successful family (*Formicidae*) of species on earth. There are somewhere between 100 trillion and 10,000 trillion ants in the world at any given moment, and one study suggested that the total weight of all ants rivals that of all humans, although these numbers are difficult to verify—and humans are also a lot more numerous, and a lot fatter, now than at any point in history.[34]

Like humans, ants specialize in the environments in which they live. They've adapted to cold climates, hot climates, high altitudes, low altitudes, deserts, forests, and, inconveniently, our homes.[35] These adaptations correspond to the fact that there are about 13,000 species of ants. In the Sahara, you'll find *Cataglyphis bicolor*, the desert ant; in the forest of South America, you might find members of the genus *Acromyrmex*, which specialize in obtaining nutrients from fungi.

Ants like *Acromyrmex balzani*—one of the group of species known as leaf-cutter ants—evolved, over the past sixty or so million years, to thrive in forest environments without depleting the resources, such as vegetation, on which they rely. Individual members of *Acromyrmex* species are genetically directed to specialized tasks, which include carrying leaves to the underground chambers their groups live in. Some defend the colony; others help to rear younger ants. Their systems are cooperative, resistant to external threats, and sustainable: they are what the late biologist Edward O. Wilson called "eusocial."

Humans also tend to specialize both as groups and within groups. Since we evolved in Africa, we've spread, over the past few hundred thousand years, to every place, however inhospitable, in the world. And yet unlike ants, which evolved into different species to thrive in new environments, we remained a single species—every person living today is a member of *Homo sapiens*, but only a small subset of ants belong to the genus *Acromyrmex*.

The reason why we remained *sapiens* lies in how information is passed between generations of ants and humans. For ants and other insects, information about how to specialize is coded genetically. They are born and soon thereafter begin their work for the colony, whatever that entails. Humans, by contrast, require a huge amount of investment in time, energy, and resources from those looking after them. Becoming a part of a culture can't be encoded in genes, and therefore learning to be a part of a culture takes time.[36]

Acculturation has the obvious drawback that very young humans aren't helpful. But it also means that any human, born anywhere, can learn how to be a member of their local culture, or to adjust to new cultures they're brought into. If you bring a leaf-cutter ant to the Sahara, it will die. If you bring a human raised in Antarctica to the Sahara, it is likely to learn the local social norms and behaviors and, eventually, thrive.

We weren't the first species to transmit information about the world socially, but we are the first to build upon that information so extensively that we colonized the planet using almost exclusively cultural, rather than genetic, adaptations. Biological differences do exist between many human groups, though. People living in high-altitude areas, such as in some parts of Tibet, are able to process air where the oxygen concentration is low. Similarly, there's evidence that people living at higher altitudes adapt to cold more easily than do others, and that people from colder regions tend to be taller.[37] Yet regardless of these differences we remain members of the same species owing to the fact that two random humans from any part of the world may reproduce successfully together.

Ants, in other words, show huge behavioral diversity for genetic reasons; humans show huge behavioral diversity for genetic and cultural reasons—and culture stopped us from speciating. But groups of all organisms, whether ants, humans, bacteria, or warthogs, specialize in obtaining resources from the environments they live in. The means of production, to borrow a term of Karl Marx's, are defined by the types of work these groups undertake for resource attainment.

So, with humans, the picture is more than just automatically adopting a genetically programmed specialty. Our specialty is acculturation, and we adopt our professions based on many factors: what we learn, what we're good at, what others need from us. Yet we also have the complex and diverse social hierarchies that lead to the inequality seen in so many cultures, including Australian Aboriginals, the Calusa, and modern Western societies. The means of production, or how we exploit local environments, combined with social hierarchies—how we each contribute to that exploitation—is a product of our diverse histories. We are the products of our relationship with our environments, social and otherwise. And that relationship has another mode: that of exploitation.

The Mode of Exploitation

A lot of focus in anthropological research is on how human groups subsist. Do they fish? Hunt? Are they agriculturalists? Pastoralists (those who care for, trade, and eat domesticated animals)? Do they trade with nearby groups, or share widely within their own groups? We aren't just what we eat—we're how we eat. It's all, it seems, about the relationship between production and consumption.

But the facet common to all is how people exploit one another. It's much better, from an evolutionary point of view, to do nothing while everyone shares their resources with you—and to use those resources to support your own family. And viewed from that Darwinian hill, we can see, and sometimes predict, how people exploit each other in different groups: the modes of exploitation of society.

Shamanism, for example, is found in most indigenous societies around the world, and although shamans don't tend to contribute much in terms of resources, they do benefit a lot from the work of others. Among the Mentawai of West Sumatra, Indonesia, people believe that shamans see invisible spirits, including those that make people sick. These spirits may be punishing people for undesirable qualities, including stinginess. Shamans are believed to understand how to rid people of these spirits—though it may be, in some cases, that the shamans themselves are the invisible rivals, not the hidden forces they contend with.

One study by Manvir Singh and Joseph Henrich showed that only one or two Mentawai shamans in a population were regularly called to heal the sick—much like people today might prefer a popular doctor working at a prestigious hospital over others.[38] And along with that prestige comes benefits: shamans are often paid with foods or sacrifices that sustain them, though they do little, if anything, to produce food themselves. While the relationship is reciprocal between healer and healed so long as the shaman performs his duties effectively, his position also comes with huge benefits that often translate to social and reproductive success. It's not easy to become a shaman in any culture, and more than 80 percent of societies worldwide require would-be shamans to abstain from food, sex, or social contact. Yet it is nonetheless a desirable position because of its high status, which, in turn, can lead to benefits that ironically include food, sex, and social contact.

Status and position in society determine how people exploit each other across cultures. In many cultures—though not all, interestingly—status determines reproductive success. Headmen among the Yanomami people of Venezuela have more wives than others, and consequently have more children. Yanomami men exchange women as commodities, and rely upon these trades when forming alliances between villages.[39] These women are exploited for sexual and reproductive benefits for men.

John Moore notes in his ethnographic review that women in many societies do most of the housework, and sometimes even most of the food gathering, though men often reap the benefits. And in gerontocracies across the world, old men, which may include shamans, contribute nothing tangible to society and yet marry the youngest women and consume many of the resources. These examples epitomize how exploitation manifests in these localities, or what Moore calls the mode of exploitation. In his 1977 paper, he wrote: "I consider exploitive institutions to constitute the evolutionary cultural core of human society . . . these institutions are independently variable through time, and are also irreversible in their progress. The synchronic arrangement of these exploitive institutions is what I would call the 'mode of exploitation' of a society, and I regard the mode of exploitation as the most significant feature for understanding the functioning of any human society."[40]

Whatever the society, and whatever the mode of production, where you have inequality in resources, status, or how either resources or status are inherited, you will have exploitation. It doesn't matter, as among Australian Aboriginals or the Calusa—or even modern society—whether people abide by the norms of how they're exploited. What matters is whether one or a few people within a group benefit from the social structures of the group more than others.[41] And it certainly matters how people are oppressed by the powerful members of the group, whether few or many. Our cultural core is not whether we fish or hunt or forage to subsist, but how our norms permit some people to hold power over others. If you want to understand a society, including your own, look at how people are exploited.

Less than Noble, More than Savage

We aren't, then, justified in calling people from pre-industrialized societies noble or savage. We find egalitarianism in some groups, like the Aché, though equality was probably much more rare in ancient groups than people such as Kropotkin would have us think. And even among groups where there is equality in resources and reproduction, as well as freedom from coercion, there are often nonetheless cultural unpleasantries that people in the Western world would reject. Before contact with people from industrialized societies, the Aché practiced infanticide at a higher rate than that seen in any culture, at 14 percent of boys and 23 percent of girls—most of whom were orphans.[42] The !Kung kill six babies per five hundred to optimize their ability to feed everyone.[43]

The Western world is in its infancy compared with even the most technologically unsophisticated of these hunter-gatherer societies. The rituals, norms, and tool-making practices we see in these groups are the consequences of the history that reaches into the core of being human—and to call those practices savage is itself a kind of savagery. Yet modern modes of exploitation are with us today just as much as ancient modes were in times past.

The idea of diverse histories—that our localities and norms coevolved, leading to hugely varying cultural practices worldwide—is darker than the simple, primitive libertarianism, stemming from Kropotkin's work, that many people continue to espouse. But it also better fits the data we see from ethnographic studies, though it makes for a messier, less palatable understanding of what makes a human a human.

More than that, though, abandoning the idea that without a state or strict social rules we would treat each other fairly can help us understand how to improve the societies in which we live now. The view that we're all, without social influences, cooperative or altruistic justifies, ironically, a lot of the exploitation that we see today. If powerful people convince those working for them that we're all cooperating toward a common goal, or that all it takes is hard work to get into a position of power, they're better able to exploit their employees for their own benefit. But if we say, instead, that exploitation is common to nearly all societies, we can legislate against the existence of positions that give people coercive power over others.

If some pre-industrialized societies, such as the Aché, create equity through community networks and sharing, people rely on their relationships to sustain them. The risk-pooling strategies we see across many cultures place much less pressure on the individual to be successful in obtaining resources, and consequently social relationships become the foundations upon which survival and reproduction occur. That does not mean that exploitative, deceptive, and even psychopathic people do not exist in these societies, or that they didn't exist until modern times. Instead, the deceptive among us in small-scale societies are likely to lack the opportunity to display their more pernicious qualities: it's impossible to be anonymous in a village or camp.

By contrast, in the sprawling, modern societies of today, we tend not to form lifelong relationships in the areas many of us move to. Isolation—seen as a huge cost in many societies, to the point that being cut off from society is a punishment or a test, as among aspiring shamans—is something many, if not all of us experience at points in our lives. And it's that removal from the social world, on which we all depend, that gives social predators, like modern-day psychopaths, their weapons against us.

When we rid ourselves of ancient, dominant alphas, we traded overt selfishness for something perhaps even darker: the ability to move through society while planning and coordinating—what some people call proactive or instrumental aggression, and which links to the broader concept of Machiavellian intelligence.[44] Psychopathy may just be the extreme and uncommon form of proactive aggression coupled with social intelligence, and often arises from awful childhood experiences. But it also gives some people the power and will to control others through deception.

Jon Ronson, the journalist and author of *The Psychopath Test*, among many other works, told me that psychopaths mimic the behaviors of those around them—an ability that lets people like Robert Hendy-Freegard appear understanding and empathetic. Academic research also shows that mirroring behaviors is one way humans form trusting relationships.[45] And yet unlike with the smaller groups within which *Homo sapiens* evolved, we live in large enough societies that someone can mirror their way into gaining trust—and then betraying it. The modern world gives psychopaths the anonymity that translates to opportunity. Machiavellian intelligence, proactive aggression, and psychopathy are therefore not the products of how we live now, but rather manifest differently today because of the modes of production and exploitation inherent to our daily lives.

People can nonetheless be said to cooperate with the psychopathic leaders of cults, just as exploited women in some hunter-gatherer groups can be said to cooperate with gerontocratic men. An airline might also thank me for cooperating when it delays my flight by six hours. But not all cooperation is equal, and just because some people are deceived or born into cooperating with exploitative practices does not mean that they aren't being exploited.

In a way, the whole philosophy of humans as fundamentally cooperative has cult-like elements, much like any bias resembling Lysenkoism. The smoke of egalitarianism—the pretense that we have a shared history of fairness—is deceptive, and it allows people in power to retain it, all because of a false promise that we have the same goals. Yet the world is messier just as our history is messier. Exploitation, just like cooperation, goes to the core of being human. But selfishness—the driver of exploitation across the living world—is much older, and much, much more difficult to eliminate.

3
The Evolution of Invisible Rivalry

Honesty is the best policy; but he who is governed by that maxim is not an honest man.

—*Richard Whately,* Detached Thoughts and Apophthegms: Extracted from Some of the Writings of Archbishop Whately *(1856)*

In Plato's *Republic,* Glaucon tells the story of Gyges, a shepherd who finds a gold ring in a cave. Gyges learns that the ring can make him invisible, and he uses his newfound power to kill the king of Lydia, marry his wife, and replace him as ruler. For Glaucon, the story represents a question about human nature. If two people—one good and one evil—discover two such rings, can we expect that the good one will continue to be good? "For all men believe in their hearts that injustice is far more profitable to the individual than justice," Glaucon says, and we shouldn't expect that anyone given the power of invisibility should remain incorruptible.[1]

Although no known technology can make someone physically undetectable, animals have an arsenal of abilities to help them exploit one another deceptively. Hiding your body isn't the only means of becoming invisible; language lets people hide in another sense: the social sense.

Yet secretive behaviors are not unique to humans—or even to animals. A hematologist I know was interested in how some kinds of cancer—leukemia, a cancer of the blood—can appear to the immune system like normal cells, allowing them to evade the body's natural defenses against invaders. The leukemic cells she was interested in seemed to be doing something more: they formed tumors at various parts of the body, and while doing so seemed to mimic the behaviors of the nearby normal cells. So, she told me, a growth made up of these leukemic cells looked and acted a whole lot like the nearby healthy tissues—with the exception that it grew uncontrollably.

Deception and mimicry are also common in the primate world. Two primatologists, Andrew Whiten and Richard Byrne, developed a catalogue of known instances of what's called tactical deception in species like chimpanzees—noting the high frequency of actions such as hiding food from alphas, the large, aggressive animals that dominate many primate hierarchies.[2] The primatologists Emil Menzel and Andrew Whiten ran a pair of experiments in which subordinate chimpanzees attempted to hide their knowledge about where food was stored from aggressive alphas.[3] In the first case, the smaller chimp, Belle, used a variety of tricks—such as running in the opposite direction from where she knew the food was—to fool the alpha, Rock. Yet for each of Belle's tricks, Rock outsmarted her and drove her away from the food she'd tried to keep.

In the second experiment, however, another subordinate, Mercury, used a different trick than any Belle had tried. He lured a local bully, Sherman, into an enclosure with a female to distract them—and then uncovered the food he'd seen stored by the experimenter, keeping it for himself. This pair of experiments suggested two seminal hypotheses: first, that non-human primates can understand what others know and use that information to attempt deceit, and second, that modes of deception are not universal among members of the same species. Animals like chimpanzees, it seemed, were capable of innovation in their trickery.

These and similar findings evaluating what is called Machiavellian intelligence drove a broader body of work that aimed to explain the evolution of human intelligence and consciousness through an arms race between deceivers and detectors—Belle and Mercury, for example. And the only difference in deception between humans and non-humans is the variety of opportunities: our complex cultures, languages, and intelligence give us different kinds of chances to exploit others without being detected. Cells, bacteria, and animals all signal information to one another—but only humans can lie.

We evolved to be our own invisible rivals: a species that evolved both to cooperate and to exploit cooperative systems for personal gain. The history of deception, however, is far older, and almost certainly predates any organism people today would call intelligent. The complexity of language gives each of us the Ring of Gyges, which can make the best or worst of our intentions invisible to anyone else—but the ability to stay hidden has been, and always will be, a feature of social life.

The Free-Rider Syndrome

A lot of contemporary discourse about cancer revolves around a central idea: the disease has probably been with living beings for as long as there has been complex life. In early 2019, *JAMA Oncology*, a leading journal on cancer, published a paper showing evidence of a tumor in a turtle that lived in the Triassic period, 240 million years ago. The tumor was a malignant osteosarcoma, a rare type of bone cancer still seen today.[4] It's unlikely that this paper gives evidence of the first cancer ever, but rather it is the oldest cancer we've found. Animals have been victims of this disease for longer than recorded history can reveal—but to understand why, we need to analyze cancer in terms of Darwinian evolution.

It's helpful to think of an organism like the human body as a unit composed of billions of cells cooperating together. Each cell, and the cells each cell divides into, relies on the survival of the host. They have a shared reproductive interest in cooperating with one another, and insofar as the body thrives, so will the cells that compose it. And yet what happens when one of those cells stops cooperating? If unchecked, that cell, and all those cells that arise from it, will flourish, and they will continue to reproduce, uncontrollably, costing the body the resources needed to sustain itself.

Cancer, viewed in this way, is an ancient traitor in our evolutionary story. At every step in our history, as we evolved from tiny, multicellular organisms into complex primates, there's been the possibility that one cell inside us will betray all the others, taking for itself while remaining a physical part of the body, eventually killing both the body and its descendants. The short-sightedness inherent to Darwinian selfishness is implied by one of the key tenets of evolutionary biology: that natural selection is, ultimately, blind. Richard Dawkins, the biologist and outspoken atheist, argued in his book *The Blind Watchmaker* in 1986 that natural selection is a slow, step-by-step process in which intricate and complex organisms are created but without an end goal. The only goal the watchmaker has is immediate: to ensure that its next replica replicates.[5]

At its most basic level, natural selection works by favoring the survival and reproduction of those organisms best suited to their environments. A primitive organism with a small number of genes, such as some kinds of viruses, might swap genetic information with its neighbor in a process called horizontal gene transfer, and if only one of those two organisms survives and reproduces, its genetic information is considered—from natural selection's point of view—superior. Imagine this happening ten, twenty, a billion times over—what you'll see at the end of the process will be unrecognizable from whatever the original form was. Natural selection, the watchmaker, wasn't planning on the final appearance. It just chose whatever form was most likely to survive at each step and moved on.

Yet problems arise when short-term benefits outweigh long-term needs. George Gaylord Simpson, an important biologist of the twentieth century, said that ninety-nine percent of all species have evolved their way to extinction.[6] This makes sense if we think of natural selection as the blind watchmaker. Blindness to all long-term consequences of evolutionary changes is a critical part of evolving, and what might help you and your children to survive may not benefit your children's children.

Cancer works in an analogous way. Strategies—insofar as a cell can have a strategy—to help itself survive deprive the body of its ability to function properly. In a sense, all the millions of years of evolution necessary for creating organisms as complex as human beings are undone the moment cancer takes hold: it breaks everything the watchmaker worked to create. And its blindness of the consequences leads to its ruin.

Cheaters Prosper (for a While)

Some writers over the past few decades have repeatedly made the same point about cancer: it represents the most basic evolutionary principle, which is that organisms must survive and multiply at any cost. As my colleague Athena Aktipis argues in her book *The Cheating Cell,* cancer is a reminder of the origins of life: take what resources you can and reproduce—and don't worry about the consequences for anyone else.[7]

Cancer is like a hyper-selfish but well-disguised person living in society, taking for themselves whatever they can and using resources uncontrollably without a care for the consequences. As a short-term strategy, just as with cancer, this has devastating consequences both for the society in which the person lives and, over time, for the person as well.

As the Dutch philosopher Bernard Mandeville pointed out three centuries ago, our private desires and natural proclivity to act self-interestedly are essential for so many of our successes as a species.[8] We create fantastically complicated ways of improving society because we compete with one another, because so many of us have a drive to stand out. The problem arises when our self-interest drives us to deceive, exploit, and harm others in the name of personal success.

A first step toward overcoming this ancient problem is to acknowledge that deception and exploitation are deeply rooted in our natures, as many ethnographic studies suggest. And yet these aren't unique to us: they are natural to every organism, and especially apparent in the case of cancer. But to devise strategies for moving forward, we should look back—and think about how our ancestors overcame the most basic forms of selfishness.

A Short History of Human Cooperation: Kinship

Lots of research—whether digital models or empirical studies with living people—shows us that humans are ready to cooperate with strangers. We don't, as Adam Smith, the economist in the Scottish Enlightenment, wrote more than two centuries ago, individually seek only to promote our own well-being.[9]

In biological theory, early mathematical models evaluating the evolution of "cooperation"—in biological terms, mutual aid—suggested that organisms shouldn't help individuals that they aren't related to or that they won't see again. This is a consequence of the foundational works of William Hamilton and Robert Axelrod, a political scientist, and—separately—Robert Trivers, a biologist, whose works on kinship and reciprocity have, together, defined the past sixty years of research into cooperation.[10]

Hamilton's and Trivers's inquiries into what are called kin selection and reciprocal altruism, respectively, defined and continue to define how cooperation makes sense from a biological point of view. Kin selection operates through the idea that humans preferentially help those with whom they have shared genetic ancestry. So given that I share (on average) 50 percent of my genes with my sister, I should favor her over any of my cousins, with whom I have a 12.5 percent genetic relationship. The Oxford biologist Alan Grafen adds that the relationship isn't necessarily absolute, but relative to the given population: if everyone in the population has a 6 percent genetic relationship to me, I should favor any first cousin with whom I have a 12.5 percent relationship.[11]

The theory of kin selection operates under the assumption that organisms act to optimize what's called their inclusive fitness—the proportion of genes identical to their own in the population. This is basically a biologist's way of saying that individuals within groups act to favor themselves over outsiders: insofar as brothers and sisters help each other, and each other's children, a greater proportion of any of their genes is likely to be found in further generations. So, when faced with a limited number of resources—for example, food or mating opportunities—and a large number of people you could possibly share with, you should, by Darwinian logic, share with your family.

Hamilton gives this as an equation, known in the literature as *Hamilton's rule:*

$$\text{benefit} * \text{relatedness} > \text{cost}$$

where benefit and cost are resources that can, at least potentially, affect reproductive success. If I'm a successful hunter and have killed a large deer, I will share some of the meat only if the benefit, adjusted by relatedness to the recipient, outweighs the cost of sharing. Suppose that if I share a bit of every deer I catch, I'll have one fewer children—which in evolutionary biology is the only currency of success. This can be understood as a reduction in my quantifiable genetic success—a 25 percent reduction, assuming I have three other children.

Yet what if, because of my ostensible generosity, my sibling who isn't a good hunter has three more children than they otherwise might have? Because I share 25 percent of my genes (on average) with my sibling's children, then a greater proportion of my genes are making it to the next generation than if I'd had just one extra child myself.

In everyday language: three nieces and nephews are of greater value, genetically speaking, than is one child. This may be why the renowned biologist J.B.S. Haldane said—apparently as a joke in a pub—that he "would lay down [his] life for two brothers or eight cousins."

Without commenting on the comedic merit of Haldane's statement, it's impossible to exaggerate how foundational Hamilton's notion is to modern evolutionary biology. And in species other than humans, it's the rule through which most instances of cooperation are seen. Yet with humans the picture is still different—and as you might have guessed, a lot more complicated.

A Short History of Human Cooperation: Reciprocity

With humans, unlike other animals, we don't normally just help kin, we also help strangers. Several years after Hamilton published his major work on inclusive fitness in *The Journal of Theoretical Biology*, Robert Trivers published one of his first major papers, "The Evolution of Reciprocal Altruism."[12]

Here Trivers speculated that people who help each other reciprocally will be at an advantage in times of need. If there's ever a time that, for whatever reason, I need wheat because my crops have failed, I'm more likely to be helped by someone I've helped in the past. In this way mutually beneficial relationships can form: two people starving at different times can have their needs met by each other, depending on who's in need and who has food to spare—the groundwork for risk-pooling relationships seen in groups like the Maasai. As with Hamilton's rule, the logic isn't complicated, but the effects are far-reaching. Individuals who can help each other and can remember who has helped them will be at a huge advantage in the game of natural selection over those who fend entirely for themselves.

With some notable exceptions like blood sharing among vampire bats and grooming among primates, most examples of mutual aid in the non-human world are among kin, though regardless we widely accept reciprocal altruism as a driving force of our cooperative behavior. Computer models that play out the success of different strategies in competitive games tell us that reciprocity is an important force in social relationships. Models of the Prisoner's Dilemma, which is a popular economic game used by biologists, are good examples.

Table 3.1. Possible outcomes in the basic Prisoner's Dilemma

We both remain loyal	You betray me, I remain loyal to you
I betray you, you remain loyal to me	We betray each other

In the Prisoner's Dilemma, the underlying idea is that the police have arrested two accomplices for a crime. They keep them apart, and, in line with a long-time tradition of mediocre television, try to persuade each of them to betray the other. If you betray your accomplice, they tell one, you'll get a reduced sentence. They tell the other the same thing, of course, trying to get them to go along with this betrayal (table 3.1).

For each of the different outcomes, there is an attached prison sentence. If the accomplices are mutually loyal, they each get three years; if they both inform on the other, they each get five years. It gets complicated, however, when one is loyal and the other is an informant. In this case, the defector gets a month, while the loyal one—often called the sucker—gets ten years. Early research into this dilemma showed convincingly that the Nash equilibrium—the mathematically predicted best strategy named after the game theorist John Nash—is to be an informant. Mutual defection with a stranger prevents the possibility of being a sucker.

The specifics of the scenario and prison sentence lengths are largely unimportant except in relation to one another: what matters is that it is better, from an evolutionary perspective, to cheat when your partner cooperates. Yet, perhaps ironically, there are a lot of times in life when we must trust that others will reciprocate our cooperative actions. Even thinking back to the wheat scenario, it's better, from a biological point of view, to take wheat when you need it, and refuse to give any when you have it.

Probably owing to its simple dynamics but important consequences, there are enough published papers on the Prisoner's Dilemma to compete with work on the Trolley Problem in philosophy (the question of whether it is ethically acceptable to kill one person to save several). Yet some stand out in the literature.

One important project conducted in the 1980s by Hamilton and Axelrod, for example, was the first to explore the dilemma from a variety of perspectives. In the co-authored work "The Evolution of Cooperation," the authors describe how they invited people from across academic disciplines to develop strategies to play against others.[13] Some engineers wrote complex algorithms for deciding when to defect and when to cooperate (that is, betray and stay loyal, respectively). Others relied on elements of chance in their computations.

Yet when Hamilton and Axelrod compared the relative success of these strategies in computer simulations, a simple one developed by a psychologist appeared to work better than the rest: tit for tat. The algorithm cooperated on its first interaction with another simulated individual and thereafter matched the strategy that its opponent last took. So if A, the tit-for-tat player, met B, who defected, A would then defect in its next meeting with B. Over time—and importantly for evolution, over generations—tit for tat turned out to be the most successful strategy.

The only major caveat of the strategy's success was that it needed to have a high enough chance of meeting the same individual more than once. This suggested an important element to the Prisoner's Dilemma that many people accept: the odds of cooperation should rise along with the likelihood of meeting the same person again. Don't bother sharing food, in other words, if you won't again meet the person asking.

A Short History of Human Cooperation: Reputation

Inclusive fitness theory and reciprocal altruism can't, however, account for an important element of human nature: we don't just help our families and we don't just help people we know. In fact, we know from empirical studies, and just from being people, that we can and do help strangers we're unlikely to ever see again. How do these acts of altruism make sense in the biologically self-interested world painted by Hamilton and Trivers?

It's impossible to answer this question without appealing to a uniquely complex feature of *Homo sapiens:* our intelligence. Robin Dunbar, a researcher who has spent his career exploring different facets of human and non-human primate intelligence, famously developed a scale for measuring a neurophysiological element against social complexity. He showed, in one of his most influential papers, that a given primate species' average neocortex size maps on to the number of individuals living in groups together.[14]

The animal with the largest body size, and an adjusted neocortex size, was—surprise!—humans, with a predicted number of social contacts of about 149. This number mapped eerily well onto ethnographic research. It revealed that this number, known in the world of anthropology as Dunbar's number, was about the average size of clans of hunter-gatherers across the world. We expect and tend to find that humans have roughly 150 social contacts they interact repeatedly with at a given lifestage, potentially satisfying the criterion of solving the Prisoner's Dilemma.

Yet over people's lives, in both industrialized and pre-industrialized society, they're likely to interact with a lot of others outside their social groups. We still enter into all kinds of relationships across cultures, and whether we're coordinating a stag hunt together or founding a tech start-up, we need to trust others we don't know well.

This is a fundamental problem in increasingly large societies: when there's an opportunity for people to exploit us in a passing social relationship, our baseline should be that they will do so. There is no reason, without context, to trust that a stranger will help you in a time of need. So how do people branch out, as they obviously do, into partnerships with 149 people?

Intelligence is, again, the answer. We began, at some point in our evolutionary story, to not just judge people from direct experience, but to watch how people treated others. This was the foundation of reputational thinking in human society and has received extensive treatment since the biologist Richard Alexander, who is partially credited with linking modern biology to ethics, discussed it explicitly in his most famous work, *The Biology of Moral Systems*.

The idea, as with direct reciprocity in the tit-for-tat model, is simple: treat others well who treat others well. It's since become known as "indirect reciprocity" and has been modeled extensively in mathematical and computer simulations.[15] It has also shown remarkable strength as a human evolutionary strategy against defectors. From indirect reciprocity, we can build to the next step of complexity: language (or more specifically, gossip).

How others transmit information about us can, as many people know from living in the twenty-first century, directly affect how we're treated. The ways we talk about people today hide the fact that the mechanism underlying the process—change to a reputation through language—is the same as it was a hundred thousand years ago.[16]

The fundamental evolutionary explanation for why you'll

find a bad Yelp rating for Walter Palmer, the dentist who lured out and killed Cecil the lion in Zimbabwe in 2015, is the same one as for why a hunter-gatherer group ostracizes someone who takes more than their fair share of prey. Our weapons of communication today are merely more technologically sophisticated, but we still feel the need to gossip and punish those we believe have done something wrong.

Along with Dunbar's number, which falls into the larger body of literature encompassing what is called the social brain hypothesis, people argue that the need for coordination drove the evolution of human intelligence because it led to the emergence of language.[17] We aren't the fastest or strongest animals, but we communicate effectively to not only hunt and forage—but to make new and better tools for doing so. In the biblical story of the Tower of Babel, when people tried to build a tower to Heaven, God changed their languages so they couldn't communicate. This removed the essential tool for coordination, and, by doing so, ruined the plan. Yet preventing people from communicating stops them not only from working together, but from talking about one another. A group of strangers who can't speak can't cooperate—or tell who best to cooperate with.

The next time someone offers the tiresome adage *It doesn't matter what you do, it's how you do it,* tell the person that it doesn't matter what you do, but what does matter is how you—or any other people—talk about it. We judge others through language, and those judgments, which enabled us to build more social and complex societies than any in the known world, rely on socially transmitted information about others. Through these mechanisms, human societies created another rule foundational to human cooperation: treat those well who have good reputations—and avoid cheating people who can tell others about you.[18]

A Short History of Human Cooperation: Signals

The emergence of language changed the ways humans could coordinate tasks such as hunting and creating tools, and also the ways they communicate information about each other. Some people argue today that this system is a difference of degree, not of kind, from communication systems seen in other animals. That is because signals pervade all social relationships across the living world—and language is just a particularly sophisticated signaling system that helps to show, and sometimes hide, our intentions.

There is a large body of literature looking at how organisms signal information to one another. Some birds display color badges known as badges of status on their chests, signaling underlying qualities such as resistance to parasites.[19] Sometimes, when researchers artificially adjust these signals, social hierarchies can change: a group of zoologists painted over one such badge, which led to the dominant individual being treated like a subordinate.[20]

Yet if it's so easy to manually manipulate these signals, why don't animals with an otherwise low social status invest, biologically speaking, in growing hairs that imitate the badges of status?

The answer relies on the idea that signals associated with any desirable quality are costly. "Costly" here does not mean a specific kind of cost, like energy or resource, but is instead defined by context. What is critical about costs is that they make a feature difficult to fake: an individual who doesn't have the underlying qualities, whatever those are, won't be able to imitate the signal effectively. This idea is known as the Handicap Principle and was developed by the biologist Amotz Zahavi in the 1970s.[21] (Although forgotten for about twenty years, Zahavi's idea was rediscovered and largely vindicated by a mathematical model developed in the 1990s.)[22] The idea, however, is simple: signals of desirable qualities are hard to fake because they represent something undetectable by sight, sound, smell, or touch, and which evolution has made costly to copy.

In the case of some birds, the underlying quality is resistance to parasites, but costly signals pervade the natural world. One example is the roar of red deer stags, which signal information about their size and weight to potential competitors and mates.[23] The roars of larger stags showed different acoustic signatures from smaller ones, and reflected the physiological limits of one mechanism for faking a desirable roar through retracting the larynx. The differentiating acoustic qualities were observable when the larynx was fully retracted—when the limits of fakery had been reached—and were linked, in larger stags, with higher reproductive success. The smaller stags cannot, or rather did not, pay the costs of growing to a physical size sufficient for producing the desirable roar (perhaps, for example, because they didn't receive sufficient nutrients at a young age).

There are many, many other cases, and you are likely to find costly signaling wherever there are interactions among organisms, whether within or among species.[24] In the case of humans, all of our communications are the products of millions of years of evolutionary pressure on how we signal information to one another. For example, it is difficult to fake even the simplest facial expressions denoting emotions like happiness or sadness—people can often tell when the expression is genuine.

This may not sound important to language. It's easy, after all, to lie about yourself and other people—but we use signals every day to form judgments about others, and not just by listening to the words they say. Think about any time you go out for a walk or buy a coffee and watch people around you. When someone walks by that you like, or maybe that you don't, you're forming your judgment using many tiny signals that you don't consciously compute: dress, walk, accent, demeanor. This ability, being an effective receiver of signals, was foundational to our evolutionary success. We formed judgments about others—and whether to trade or hunt with them—based on signals, many of which were almost undetectable.

When someone lied, as they did and do, about whether they would be a cooperative partner in hunting, a good compatriot in battle, or a loyal friend, humans had to become efficient at detecting deception. And research around deception detection has yielded mixed results: studies suggest we can sense lies a little over half to two-thirds of the time.[25] That's obviously not perfect—but it might have been good enough that, over time, people could weed out overt liars, maximizing the odds of working with cooperatively minded others. The costs of signaling are still there much as they are with other animals, but they reflect the possibility of punishment and ostracism when caught trying to deceive others.

A Short History of Human Cooperation: The Social Market

All these factors—kinship, reciprocity, reputation, and language—fit together to help explain how and why we cooperate with others, even others we may not know. And each is implicit in the general picture painted by the biologist Mary Jane West-Eberhard several decades ago: the notion of social selection.[26]

Natural, sexual, and kin selection explain the evolution of organisms through survival, reproduction, and helping genetic relatives (kin), respectively. Together, they maximize a given individual's inclusive fitness or genetic success. Social selection, however, is an essential part of group living where individuals can choose partners, whether for trade, hunting, mating, or any other behavior that requires cooperation—and trust.

Think, for illustration, of people exploring a market. It's not an ordinary market where people are looking for goods. They're looking for partners: they need someone to (for example) hunt with and the people at the market stalls are trying to advertise—or, perhaps more carefully put, to signal—themselves as good candidates.

As people walk through, hypothetically speaking, those at the market stalls yell about their excellent hunting techniques or how generous they are when they catch game. These exclamations inform our decisions, though we rely on our reception of the signals they display when making our judgments. We want to know if the vendors are trustworthy.

This is biological market theory.[27] People have an incentive to advertise themselves as good partners for cooperation or for mating, and these advertisements are important factors for the creation of human coalitions. We select, socially speaking, those we believe will work best with us, relying implicitly on all the factors essential to the evolution of cooperation. *Are they related to me? Will they be generous if I am generous to them? Do others speak well of them?* Each of these questions is part of the marketing game, and language makes it possible to describe ourselves positively—or others negatively—in ways likely to benefit us in the Darwinian game of life.

The Consequences: Survival of the Nicest

Taken together, this short history of human cooperation depicts an optimistic scene: one where, despite the proclivities for exploitation and competition natural to all animals, language evolved among humans to allow us to sort those who are trustworthy from those who are not. Because of our ability to punish and ostracize people who don't signal their cooperativeness or who try to lie with false signals, we created a circumstance unique to all living creatures: we self-domesticated. Much as humans artificially selected for the dogs with the best temperaments, humans selected for those who were the most cooperative.

This view, dubbed "survival of the friendliest" by many in the evolutionary sciences, joins all the fundamentals of human cooperation.[28] For Richard Wrangham, the primatologist who believes we selected against reactive aggression—the kind of aggressive behavior that stems from emotion or a response to threats—the social complexities that led *Homo sapiens*, relying on kinship, reciprocity, and reputation, to form cooperative groups also allowed us to unite against the tyranny of dominants we inherited from our evolutionary ancestors.[29]

The forceful, coercive dominants no longer had the upper hand, from this perspective. Coalitions of weaker individuals did. And together with the weaponry hominids started to develop over the past million or so years, we had not only the tools but also the social and linguistic sophistication to overthrow aggressive alpha individuals. We became over time kinder, more cooperative people: the same people that use language to continue to find cooperative partners, maximizing their own and others' inclusive fitness—fitting the earliest notions of morality onto Darwinian foundations.

Some people who study how culture evolves in an analogous process to biological evolution take this argument even further. They argue that groups of cooperative people who worked together effectively outcompeted less cohesive groups. This process, known as cultural group selection, led to the success of cultures that championed collaboration, cooperation, and cohesiveness above individual-level selfishness.[30] Using punishment as the ultimate deterrent against free riding, groups instilled, over time, teachings that internal cooperation is best. Cultural group selection implies, to use the language of this body of work, the internalization of norms that promote cooperation within the group. Group-level competition, rather than subversion or exploitation from within, then becomes the primary antagonist in this narrative of human history.

The Problem of Opportunity

What we don't account for, in this rosy view, is the implications of the collective punishment process. It's undeniable that people tend to do well when we live and work together; we aren't therefore going against our self-interest—as distinct from selfishness, which implies inflicting a cost on others—by cooperating. We pool risk and spread resources where others are in need. We don't, at least in many democracies, have aggressive dominants controlling our behaviors—though people living in autocratic states cannot say the same. Nonetheless, even where everyone has a political voice, there are many ways in which the mechanisms underpinning cooperation can be, and are, exploited.

In modern society, for example, we have a term for how kin selection interferes with cooperative systems: nepotism. When politicians or other high-powered people in industries favor their own family, placing their interests above others, we don't ignore their preferential treatment because of evolutionarily instilled motivations to help kin. We say, instead, that their children have gotten an unfair advantage.

If the only reason to cooperate is that we're less likely to be punished for defecting or less likely to be ostracized from cooperative systems in the future, shouldn't we expect defection in situations where these risks aren't high? And don't we see that pattern when there are power differences in relationships? When an employer mistreats an employee within the boundaries of legality, can the employee punish the employer? Can a poor country punish a rich one for mistreating its citizens? Generally, the answer to both is no.

Elon Musk, the Tesla (and more recently, X) CEO and sometimes richest person in the world, famously took U.S. government subsidies to fund his entrepreneurial ventures. And yet when faced with paying taxes on his extreme wealth, he complained; in 2018, he paid $0 in federal income taxes, according to *ProPublica*.[31] The state cooperated with him, and he defected (to use the language of the Prisoner's Dilemma) from the state—a real-life, albeit inexact approximation of this hypothetical scenario.

More broadly, there is generally good evidence that people can and do fake cooperative signals for personal benefit. Virtue signaling—showing off your moral inclinations so that others think better of you—is one example. When Russia invaded Ukraine in early 2022, for instance, centers for the arts across the world canceled productions of Russian operas, and Netflix, the U.S.-based streaming giant, ended production of a contemporary, Russia-set series based on Tolstoy's *Anna Karenina*.[32] It is unclear, however, how boycotting Russian cultural exports from the nineteenth century helps Ukraine to win the war. And given the anti-authoritarian themes frequently found in Russian works of art, sending costless signals of commitment to the Ukrainian cause doesn't appear to do much other than boost the reputations of the signalers.

A group of researchers based in Ireland, Spain, and England investigated how well signals predict the likelihood of following through on proclamations of virtue. They found that people who advertise their intentions to donate to charities for public adulation were less likely to actually make donations.[33]

They were less likely, to use the language of Zahavi's signaling theory, to pay the costs that would have demonstrated the veracity of their signals.

Awareness of costless displays of virtue can undermine the movements the signals ostensibly support. If someone is enraged by something on social media, such as police brutality in the United States, and becomes a known voice of support for that movement, and then turns away when something else becomes more interesting to the person, that may dissuade others from taking the first issue seriously.[34] Insofar as people believe that known figures in a movement are interested only in self-promotion, the movement can be eroded. Following through on our words is important not only for us, but for the issues we believe in.

The fundamental problem, however, is that the person who exploits cooperative systems successfully will be better off than those who don't. If I can defect in a cooperative relationship or favor my family unfairly without others knowing or caring—and who continue to cooperate with me, moreover—won't I be better off than everyone else?

This is the problem of opportunity. If people are presented with an opportunity to cheat, many of them will. And it's likely that there will be little anyone can do to stop or punish them.[35]

The Problem of Intention

Virtue signaling raises an important question we all spend a lot of time worrying about, sometimes without realizing it: who is really committed to the causes they espouse? The authors of an article in the *New York Times* in 2019 argue that we can't, and possibly shouldn't, separate genuine believers from fakers among people who signal virtue.[36] We want to, but can't always, or even often, tell who the "devout actors" among us are—to use the language of some contemporary anthropologists.

In fact, some argue that an important, if not essential, human behavior is to imitate people we believe to be committed to causes. This is because, given that appearing committed to moral causes is likely to bring reputational benefits, appearing to be like those committed to those movements may also bring those benefits. Being seen as a principled person is good for you—and even when some people deviate from their principles, research shows they are effective at explaining how their behaviors nonetheless conform to what they (appear to) believe.[37]

The problem of reading intention is the foundation of the problem of knowing whom to trust, and is the characteristic that makes us essentially different from any other species. When a dog bares its teeth, possibly indicating its intention to attack, we can tell whether its signal was honest by whether it attacks. With people, it's much harder to tell whether signals are empty. They can lie about their intentions, and our own biases affect whether we believe them. Coupled with the problem of opportunity, this problem of intention suggests that we can't rely on computer models to tell us when a person really wants to help others or when they want to cheat, but just lack the opportunity to do so.

In the barest interpretation, social norms can be seen as a collection of rules that prevent free riding and exploitation. But who is to say that most people follow those rules only for as long as they won't get caught?

Language and reciprocity together make cooperative systems more complex and widespread than any without these mechanisms. The social brain hypothesis effectively describes the supreme importance of cooperation for the evolution of intelligence. We are the most intelligent because we are the most cooperative, and vice versa. And because of these attributes, we evaded the fate of some other less intelligent primates—those still at the mercy of overtly selfish dominants.

We didn't, following the self-domestication hypothesis, only select against reactive aggression, we selected—however unintentionally—for the people best able to mask their selfish intentions, and who planned their aggressive tendencies the most effectively: we selected for what is sometimes called proactive aggression.[38] The popular and comforting notion of "survival of the nicest" isn't quite right: it's survival of the opportunist, who is sometimes nice, and sometimes not.

The social brain hypothesis has another element. According to Nicholas Humphrey's influential paper from the 1970s, from which the term "Machiavellian intelligence" was developed, the social conditions under which primates evolve selected for those best able to navigate their groups for Darwinian benefit.[39] As much as cooperation and group living helped to enhance individual powers, so too did individuals need to maneuver within groups to obtain better mates and better coalitions.

Invisible rivalry is a step more complicated. As much as we evolved to use language effectively to work together, to overthrow those brutish and nasty dominants that pervaded ancient society, we also can (and do) use language to create opportunities that benefit us, as many modern tales, like that of Dan Price, show. We use language to keep our plans invisible.

Humans, more than other known organisms, can cooperate until we imagine a way to compete, exploit, or coerce, and almost always rely on language to do so. All the novel strategies of Belle, Rock, and Mercury are few compared with those humans could use to hide food when communicating linguistically. We use tactics to engender trust, to win the adulation of others, to make ourselves rich through cooperative ventures like starting companies, and to give ourselves an opportunity to cease cooperating. This is invisible rivalry—and the worst effect of it is that we often reward those in our societies who are the best at it.

Invisible Rivalry

Many who work in the broad area of human evolutionary studies rightly argue that we are uniquely ultrasocial: we are supercooperators. We build complex societies and rich cultures: think of all the customs and practices across the world, and all the archaeological evidence of those complex cultures that came before ours. In the past several thousand years alone, we've built the wonders of the ancient world, developed far-reaching religions that link hundreds of millions of people together, and invented intricate technologies that no person could ever do alone. We couldn't have done any of this without intense cooperation, and the fact that civilization hasn't (yet) collapsed is proof that cooperative societies can endure.

Although supercooperation is a fundamental attribute of human nature, it's only half the picture: the cleverest people among us are capable of subverting cooperative practices for their own benefit—much as Glaucon predicts, using the allegory of a ring, in the *Republic*. Language is our ring of invisibility: we don't know who these people are because the best of them will evade detection and hide their exploitative practices from society. As Niccolò Machiavelli put it in 1532: *Ognuno vede quel che tu pari, pochi sentono quel che tu sei* (Everyone sees what you appear to be, few experience what you really are).

Invisible rivalry brings together the problems of opportunity and intention. Intelligent exploiters are the invisible rivals in our societies—following the rules until they no longer need to. They fake it till they make it in society—or perhaps, as with a cancer inside its host, until they break it.

Every system, whether political, religious, social, or industrial, is open to exploitation, and people will always find a way to exploit. This is analogous to how cancers and viruses hijack the body for their own purposes: to survive, multiply, and spread. And when we think we've outmaneuvered the disease, it comes back with some novel response we aren't expecting. With the invisible rivals in our own societies, as with disease, they are often too subtle in strategy or too devious in character to be eliminated.

Even the tools we use to expose exploiters are open to exploitation: think about how people across the political sphere accuse others of virtue signaling or abusing a well-intentioned political movement for their own gain. Think about how many times over the last millennium people with religious power have used their influence to enrich themselves and their families. Or think about how often in modern times leaders are accused of taking enormous monetary sums to promote some industry. Most of the time, these people are already in power and our complaints against them come too late, because our understanding comes too late.

For opportunistic exploiters, social barriers—norms, laws, rituals—function as guides about what they cannot do, and force them to invent novel methods of subverting the system for their own benefit. This is true in an analogous sense across the biological sphere, from the cellular level to that of human society, and the only differences are a consequence of the environments in which the exploiters operate. The ways, then, that cancer cells exploit the natural defenses of the human body, that a chimpanzee tricks others about the location of food, and that some clever people subvert the laws and norms of their own communities are all ultimately products of the same fundamental evolutionary principle: if a system can be exploited, it will be.

This is the core view upon which the rest of this book will rely, and through which we can better understand almost any social behavior. We're all supercooperators and supercompetitors. But that doesn't mean exploitation will always win—just that we should view human action through a critical lens. But there's much more to do to confront it.

4

Capital

In 1925, Marcel Mauss, a French sociologist, published a monograph titled *The Gift*. In it, he explored how the universal practice of gift-giving among humans drives relationships within, and between, cultures across the world. The Maori people, Mauss noted, place a peculiar importance on reciprocity, specifically around gifts. For the Maori, gifts aren't just objects passed between people. They retain a spiritual force that stays with a person regardless of what happens to the object. Mauss gives a description from a Maori informant: "Suppose you have some particular object, taonga, and you give it to me; you give it to me without a price.... Now I give this thing to a third person who after a time decides to give me something in repayment for it (utu), and he makes me a present of something (taonga). Now this taonga I received from him is the spirit (hau) of the taonga I received from you and which I passed on to him. The taonga which I receive on account of the taonga that came from you, I must return to you. If I were to keep this second taonga for myself I might become ill or even die. Such is hau ... the hau of the taonga."[1]

The second gift in this scenario is the spirit that accompanies the object, not the object itself, which is merely the means through which a gift is given or repaid. And that spirit, the Maori believe, can kill someone who doesn't repay a gift received.

Biological accounts of cooperation with non-kin rely at their foundation on reciprocity, and specifically reciprocal relationships between two individuals, grouped together under the term "dyads" in the academic literature. We see dyadic relationships across the spheres of life, from amoebas to animals with culture, including humans. The manifestations of that foundation—reciprocity in dyads—are driven by the social and ecological circumstances organisms live in, and are colored in humans by the nebulous umbrellas of ritual, norms, and culture.

The diverse histories of human cultures then rely, ironically, on the universal principles of cooperation around which they appear to have evolved. But societies nonetheless share the fact that they have norms around reciprocity, gift giving, and cooperation, without which trust would disappear and no culture could survive. Norms govern how we treat each other, and without this knowledge, transmitted down and evolving, generation after generation, we'd lose the shared social qualities that distinguish us from the rest of the natural world.

The centrality of these norms to human cooperation is exemplified in a simple formulation: "Treat others as you would like to be treated."[2] This is known colloquially as the Golden Rule, and as the late philosopher Derek Parfit pointed out, the rule has appeared in the writings of major cultures across the world—some of which may not have had any contact. It is a fundamental idea to many societies' ethical framework, and it is no coincidence that it coincides almost exactly with evolutionary ideas of how reciprocity forms the groundwork of cooperation.[3]

Our diverse histories determined how people interpreted the rule in their own languages, which reflected the cultures through which these norms evolved. More than two centuries ago, the philosopher David Hume likened local ethical norms to the paths two rivers take. The Rhone and the Rhine, he noted, have peculiarities that result from the grounds on which they run, not the mountain from which they spring.[4]

The story of the Golden Rule, as with the stories of many of our norms that tie to ethics more broadly, is like that of the Rhone and the Rhine. The source of the different formulations of the rule is like the mountain, which is the steadfast principle, discovered by Darwin and his students, from which our norms derive.[5]

Everywhere, across the world, rituals, beliefs, and norms around gift giving are just the local manifestations of that principle. But it's culture, and how it evolves, that gives us the rules that govern everyday life—and the means of using it to maximize what matters to all organisms: capital.

From the Top

Capital has a lot of associations for people interested in theories of political organization that govern economic activity. But the word "capital" has a long history, deriving from the Latin *caput*, or head. It's been used over the past two thousand or so years to describe the part of a body that is above the others, something in control that confers or houses power. In Middle English, *capital*—referring to "the top"—was used to speak about heads of cattle, which were an important form of wealth.

In the eighteenth century, Adam Smith, credited with founding economics, defined capital more broadly as "that part of man's stock which he expects to afford him revenue."[6] Today, many people use "capital" to refer to the resources a person controls, whether those resources are assets, like a house, stocks, or a painting, or just liquid money. The broader notion of capitalism implies a system of institutions defined by the aim of maximizing those resources: insofar as everyone pursues self-interested maximization of wealth, it's likely that everyone, on average, will be better off.

Traditionally, people have seen capitalism as an alternative economic system to, if not the opposite of, socialism, which focuses on the means of creating and distributing resources among members of a society. This is fundamental to Marxist thinking—but it's interesting to note that no ostensibly socialist state has succeeded in effectively redistributing resources, and further that, as long ago as the sixteenth century, the Dutch thinker Erasmus argued that no political system could overcome human greed, or *Plutus,* father of folly in his *In Praise of Folly*.[7]

In anthropology, the concept of capital is broader still than in economics, and while it is often used only to reference forms of economic wealth, ethnographers often show how, across cultures, forms of capital differ in importance and measurement. But for anthropologists, whether and how capital is transmitted between generations is the critical point for understanding inequality and exploitation. Viewed from above, and accounting for our diverse histories, we can start to understand the force of Erasmus's claim—and how invisible rivalry allows some people to exploit their social systems to maximize capital.

Resource Capital

When most people think of capital, they are probably thinking of resource capital: the wealth, or the potential for wealth, a person owns or controls. In many Western democracies, this is usually straightforward: your resource capital is the money you have and the assets you control, though Marxists use the term only in reference to money that allows for the accumulation of wealth. That accumulation requires a surplus—there is more wealth, and are more resources, circulating than a person needs to subsist—which in turn permits control over the means of production. Resource capital gives people power over those who can't sell anything but their own bodies for working.

Humans aren't, however, the only species to control resources. Some birds, including species of songbirds such as the British hermit thrush, aggressively control territories during mating seasons.[8] The same noises and songs we find beautiful may attract mates or convey aggression to other males intruding in these territories, and to the degree that a male can defend a territory, he has control of resources that others do not. What's more, some birds, like the acorn woodpecker, inherit territory from their parents, fueling inequality: about one-quarter of a male acorn woodpecker's land comes from its parents.[9]

This control of resources drives not only the mating and reproductive success of the male, but also the mating patterns more broadly in the population. If, for example, a male can aggressively control a large or resource-abundant enough territory, he can mate with several females and provide food for the offspring of each. Females may then choose an already mated male with a large territory over an unmated male who can't provide as much. This is known in ecology as the polygyny threshold, and refers to any situation where a female is better off, in terms of provisioning her offspring, mating with a male who isn't, to use a colloquial term, single.[10]

The polygyny threshold refers, in its original formulation, to bird populations, but it extends to all sexually reproducing species where resource capital is critical, which it always is. It also suggests that the American actress Scarlett Johansson wasn't accounting for resource control when she said in 2006, "I don't think human beings are monogamous creatures by nature."[11] Where there's more resource equality, monogamy does make sense: no one has anything to gain by mating with an already-mated partner. Polygyny—or polygamy, which refers to both sexes—makes sense only when we reach a degree of wealth inequality that skews reproductive success toward those who have a lot of resources.

We see this in mating patterns in countries such as the United States today, where we've (arguably) reached the polygyny threshold for some people—particularly given our extreme reliance on inheritance for resource capital. (Polygyny, in this case, refers not to marriage, but mating—married people do, of course, mate with people they aren't married to.)

The modern advent of billionaires and their ostensible altruism is a good example of the polygyny threshold being reached—and for me, this highlights invisible rivalry in action. Over the last several years, some of the world's most wealthy people have pledged much of their capital to philanthropic causes. Jeff Bezos, the founder of Amazon, said in 2022 that he would give most of his wealth away during his lifetime. And others, such as Elon Musk and Sam Bankman-Fried—the disgraced ex-CEO of FTX turned convict—have pledged to support new-age charitable movements, such as Effective Altruism.[12]

Yet in evolutionary thinking, these pledges don't entail meaningful costs. Signaling theory suggests that costs don't necessarily refer to economic costs, but rather the impact to an individual's genetic success. Even today, this remains the currency of human biology.

Biologists define altruism, in turn, by whether a behavior involves paying costs.[13] If I give away enough money that I won't be able to pay for childcare and consequently cannot afford children, I'm being altruistic, according to evolutionary biology. But if I give some change to a homeless person on the street, the cost is negligible, and "altruism" isn't an appropriate word.

In the modern world, wealth plays into this: many people, particularly in the middle class, wait to have children until they feel more financially stable—and in fact Charles Darwin argued in *The Descent of Man* that this is what rational people should do.[14] Evolutionary research over the past few decades supports this trend, and the rising cost of homes and childcare are, some people argue, affecting birth rates. Data from the U.K.-based charity Pregnant Then Screwed, for example, suggest that nearly one-third of new parents in the United Kingdom cannot afford another child because of high and rising childcare costs.[15]

Whether giving is altruism is consequently relative. A hundred pounds to a corporate lawyer isn't the same as a hundred pounds to someone living in an urban slum in Mumbai—and certainly not the same as a hundred pounds to a billionaire.

The dark side to this false altruism is that, because of their resource capital, rich people can afford to have more children in high-income societies (billionaires, in some notable cases, have nine or more). And some, such as Elon Musk, apparently believe that it's their duty to spread what they think are their superior genes throughout the world.[16]

His resource capital permits this, much as the social power of a warlord in the past permitted high reproductive success. Population genetic data suggest that as many as 0.5 percent of males worldwide today may descend directly from Genghis Khan, the thirteenth-century conqueror—though instead of an army of Mongol warriors, billionaires today have resource capital.[17] The number of offspring will be less extreme than what was boasted by a warlord who lived eight hundred years ago, but it looks likely that an exaggerated proportion of future generations will call the billionaires of today their ancestors.

Despite promises of giving, and many of the world's richest people signing up to the Giving Pledge—a movement in which ultra-wealthy people promise most of their wealth to charity by their deaths—very few, if any, have paid the costs of biological altruism. Resource capital, and its transmission between generations, translates to reproductive inequality. And yet, because the rich—and those working for them—use language effectively, they can maintain a good public image while exploiting society for their biologically selfish benefit. They can remain invisible.

Social Capital

Relatedly, altruism in the colloquial sense—sometimes called psychological altruism, and which is characterized by good intentions—has another function, which many people with resource capital look to capitalize on: it allows the exchange of wealth for improved social standing. As with everything in anthropology and biology, the consequences of these exchanges are contextual. There's a well-known scene in *Curb Your Enthusiasm*, for instance, where the central character, Larry, attends a charity event. He's made a donation to a nonprofit that earns him a building wing with a placard that reads "Wing donated by Larry David." He's happy with this, until he sees a second wing that reads "Wing donated by Anonymous."

"It just looks like I did mine for the credit," he says in his idiosyncratic way that fans may remember from the show's better days. "As opposed to, you know, Mr. Wonderful Anonymous."

The Mr. (not so) Anonymous in question—the actor Ted Danson, we learn—capitalized on his donation more effectively, drawing attention to his generosity without advertising it openly. Larry looks worse only by comparison, and his social standing doesn't get the boost he hoped for when making his donation.

From the view of the economic or anthropological sciences, Larry and Ted were competing for social capital. This is a more nebulous term than resource wealth, which usually involves observable goods like land or livestock, but social capital nonetheless drives a huge amount of human behavior, and arguably is more important in determining a person's livelihood and longevity than is resource capital.

In a biological market, social capital drives social selection. Mathematical and computer models, as well as laboratory studies, consistently show that individuals with good social standing are likely to be chosen as partners for cooperative tasks.[18] And people almost always take a person's reputation into account when deciding to form a partnership with them—and many relationships end when a person's reputation is damaged.

In 2023, the shoe company Adidas dropped a lucrative partnership with the American rapper Kanye West, reportedly because of anti-Semitic comments he made and tropes he posted on social media. The company probably made this calculated decision to maintain its social capital by—at least temporarily—giving up a huge amount of potential resource capital.[19]

Yet reputation matters to us all, not just public figures. Our social capital is both what people think of us, in the broad sense, and our relational wealth: the influence, resources, and status of the people to whom we're connected. In pre-industrialized societies where resource storage and inheritance are less common than in large capitalist economies, transmitting social capital can be essential for guaranteeing the success of your children.

The cultural anthropologist Polly Wiessner showed that, among the Ju/hoansi bushmen of Botswana, children inherit not only land rights from their parents, but cooperative partners for exchanges of goods.[20] There is also evidence that, among the Lamalera hunter-gatherers of Indonesia, children can inherit both property and important positions from their fathers.[21] And while osotua systems may largely benefit Maasai society, it'd be implausible to assume that all resource-pooling relationships are equally beneficial. There will be some people who just can't contribute as much as others—and there will always be people who free ride when they think they can get away with it.

The social status of people's relatives can also benefit them. In western China, Buddhist people sometimes send sons to monasteries, which can effectively block successful reproduction: these monks are not allowed to have sex or marry. But research suggests that this practice doesn't damage the parents' genetic success: grandparents with a monk son have more grandchildren, on average, than do others. Having a monk for a son, in this case, is adaptive, to use the language of biology.[22]

The evolutionary theorist Richard Alexander argued that similar patterns hold across societies, no matter how large or sophisticated. For example, some people spend a lot of time—and interestingly, a lot of money—trying to prove that they are related to war heroes who sacrificed their lives for their fellow soldiers. Giving up anything, from one's money to one's life, can be interpreted as investment in social capital, which translates to greater relational wealth for one's family.[23]

Social capital can be a network of relationships, reputation, a heritable position like a dukedom, or a culturally defined grouping such as a caste. When we meet a stranger, we pick up on signals about these qualities, consciously or otherwise, which give us information that we use to make social decisions. A person's accent tells us about their social identity, for instance, and studies in the United Kingdom have repeatedly shown that accent is a critical driver of professional success. People whose accents are associated with a high social class in England have a better chance of getting competitive jobs than do people with vocal patterns that suggest a less privileged background.[24] Accent analysis is even used in criminal cases, with some phoneticians like my colleague Francis Nolan working in a forensic capacity, and helping to determine a speaker's identity when, for example, someone allegedly threatened to commit a crime over the phone.[25]

Today, we see the implications of social capital all over society. When Donald Trump gave his son-in-law, Jared Kushner—who had no political experience—a job in the White House, both were capitalizing on the social capital of Trump's position as president. And in 2022, major media outlets started reporting on "nepo babies"—people who were related to celebrities and who benefited from their family's social capital. The lists of these beneficiaries are long and sometimes surprising, and often leave the feeling that there's little a person can do to succeed without family connections.[26]

This isn't a new problem, though. Nepotism is a benefit for some, and a hurdle for others, that derives from the importance of social capital over our evolutionary history. We use our connections, familial or otherwise, to improve our livelihoods, and thereby convert the social capital we win or inherit into resource capital. And that's unlikely to change, no matter how many donations, ostensibly anonymous or not, wealthy people make.

Embodied Capital

The final type of capital anthropologists and ecologists focus on is limited to the body, though it determines how that body interacts with its environment: embodied capital. This is the only type you will find across species, because regardless of whether an organism reproduces sexually or relies on resources it can control, how well its body interacts with the environment determines whether it will survive long enough to reproduce.

This is true whether we're talking about a human, chimpanzee, triceratops, or fungus. Embodied capital permits the organism to obtain the resources it needs to survive and evade predation. Yet in a sexually reproducing species like a lion—or any mammal—embodied capital also determines the individual's likelihood of reproducing successfully.

This is because of a special subtype of social selection called sexual selection, which Darwin discussed at length in *The Descent of Man*. Animals (humans included) need to attract mates to reproduce, and to do so they need to display the qualities that mates are likely to find desirable. This differs from species to species, and among humans, from culture to culture, but the pattern always relies on a fundamental rule: if you don't reproduce, it doesn't matter if you survive.[27]

One problem, however, is how organisms can best advertise the hidden qualities they have. Muscles are observable, but what about strong resistance to parasites? Non-human animals, including some birds, have evolved signals that conspecifics (members of the same species) can read. Some researchers have suggested that a male's bright plumage is a signal to females that shows high resistance to parasites, and females use this information to make mating decisions.[28] Because the bright plumage is costly—it may increase the risk of predation or just require a lot of energy to grow—birds without the strong resistance won't be able to develop it.

Insects also rely on signals to attract mates—though not all these signals are reliable. Some flies and spiders bring nuptial gifts to potential mates to show their prowess as food gatherers or hunters. The females may accept these gifts—which, strangely enough, tend to be wrapped—and mate with the giver. Yet because what's inside the wrapping isn't visible, some males, spider and fly, put whatever they can find, edible or otherwise, in the gift box. In this way, they falsify their signal, convincing

Capital

Greater sage grouse lek near Bodie,
California (Jeannie Stafford/USFS)

the female to mate with them without having to find or share any food.[29]

The males of some species even group together during mating seasons to display their relative embodied capital and attract mates, in a process known as lekking. Blackbucks and greater sage grouse are notable examples: males gather in large, open spaces and stand like statues, waiting for females to select them—or not.[30] Yet humans also have mating rituals that resemble lekking. The Wodaabe people of Niger organize an annual competition called the Guérewol, in which men stand, dance, and sing, sometimes for long periods.[31] During the competition, unmarried women choose a man they find the most attractive.

Darwin recognized the fundamental importance of mate selection in human evolution, and how human mating practices mirror those we see in the animal world. Embodied cap-

The Guérewol festival of the Wodaabe tribe
(Dan Lundberg, CC BY-SA 2.0)

ital is historically and today an important quality in mating: heterosexual women in Western cultures often rate height, for example, as an important quality in a potential partner.[32]

Yet because of the diversity of human culture, the subtypes of embodied capital that matter to prospective mates—and moreover, to survival and caring for offspring—vary from place to place.[33] Among hunter-gatherers, hunting skill is an important quality that often determines a person's reproductive success. In 2003, anthropologists published a study of the Meriam people, who are islanders inhabiting the Torres Strait off the coast of Australia. The Meriam are renowned for turtle hunting, for which they organize hunting boats throughout the year.[34]

Although the hunts rely on people operating and steering the boat and younger men who jump on the turtles they chase, the term "hunter" (*ariemer le*) is reserved for the leaders only, who are older men. These leaders direct the boats, while younger apprentices serve as jumpers (*arpeir le*). Hunting is an important affair among the Meriam. Although men who are successful at hunting don't eat more than others, they organize large feasts for their community.

The 2003 study found that hunting prowess was linked to reproductive success. Specifically, good hunters started reproducing earlier, had more reproductive partners, and tended to mate with partners who were considered high quality in the community. The authors interpreted these results to support what's called the costly signaling hypothesis: hunting prowess is a signal that people can't fake, and potential partners prefer potential mates who are known as better hunters.

Embodied capital doesn't just involve physical skills, though. The knowledge a person possesses—about agriculture, cattle herding, or even religion and technology—is also included. Cultural information must be learned, and sometimes takes people years or decades to master before they can become productive members of their society.

This relates to a mystery in anthropology around what's called the demographic transition. We'd expect that as food becomes more widely available, people should have more children. Yet after the industrial revolution of the eighteenth and nineteenth centuries, those living in modernizing societies started having fewer children, not more, on average.[35]

Several explanations have been proposed for this finding, but one involves embodied capital. Once a society becomes sufficiently large and technologically sophisticated, parents need to spend more time and resources investing in their children's education. More children are better from a Darwinian view in a vacuum, but the costs of teaching your child to be a member of society—and one, moreover, that other people in society will find a desirable partner, romantic or otherwise—can weigh in favor of fewer children with more embodied capital.

Some researchers even argue that the lengthy period of human childhood and adolescence—eighteen to twenty-one years, compared with fourteen or fifteen in chimpanzees—is a product of the sheer amount of information a person needs to absorb before successfully starting a family.[36]

Contrast this with the life of a termite, which hatches and immediately starts its genetically programmed work, whether that involves colony defense or resource acquisition. Human culture is too complex and variable for analogous programming, and we therefore require huge amounts of training, socializing, and play to start effectively interacting with our environments. This is just one of a few competing hypotheses, but it aligns with modern research suggesting that many people in Western cultures just can't afford to have more than one child.

Capital in Context

Broadly speaking, our cultural histories shape our environments, but our environments also shape our cultural histories. A study by Toman Barshai, Dieter Lukas, and Andreas Pondorfer in 2021 aimed to determine just how much, and compared foraging, reproductive, and social behaviors in more than three hundred species, including several hunter-gatherer human groups, mammals, and birds. Their findings suggested that culture, at least historically, doesn't make us as different from nonhumans as we might like: in general, where people ate fish, local animals ate fish, for example.[37]

It gets weirder, though. They found that, in places where humans hoard food, animals tend to do so, too. Males of many species tend to monopolize mating with females in similar ecological environments. And perhaps strangest of all: in environments where humans tend to have social classes stratifying society, both mammals and birds tend to have dominance hierarchies with only a small subset of individuals within groups reproducing successfully. Others are demoted to subordinates, and help with the reproduction of the dominants.

Just like other animals, we are the products of our physical environments. And because those environments differ so much and our cultures evolve so much faster than anything biological, the ways people build capital and compete differ dramatically across the world. Capital is defined by cultural and physical context.

It's becoming increasingly clear just how forceful that connection is. In 2010, a group of anthropologists interested in how inequality develops explored the contextuality of capital explicitly in hunter-gatherer, horticultural, pastoral, and agricultural groups of people.[38] They identified more than forty types of wealth across the societies, which included hunting returns, digging skill, body weight, exchange partners, land, and cattle, among others—forms of wealth that, while different, fit under one of the three umbrellas of capital.

The forms of wealth in each culture mirrored the forms of subsistence, supporting the idea that our environments are a crucial part of our diverse histories. But how those forms of wealth are transmitted between people, and from generation to generation, is no less important, and is the product of culture, environment, and the individual struggle for power.

Some societies have dowries, where the father of the bride transfers wealth to the groom. Others have a bridewealth, where the groom or his family make a payment to the bride. (The former of these is linked with dowry-murders in modern India; the latter of these, according to recent research, is associated with oppression of women across the African continent.)[39] In some societies, inheritance of social or resource capital is through the father, while in others, the mother; these are termed patrilineality and matrilineality, respectively.[40]

The researchers interested in inequality captured transmission of wealth as a single statistical variable, which they compared across the cultures they investigated. They found that resource capital is more frequently inherited than the social or embodied varieties, and that inequality is greatest with resource wealth across societies. What's more, agricultural and pastoral societies relied on resource wealth more than the others, showing how increased reliance on ownership, rather than embodied capital or social relationships, drives inequality as groups grow larger. As people inherit more, they keep a greater proportion of society's wealth for themselves. Equality declines, classes can form, and the poor start to rely on the rich. More wealth inheritance allows for more exploitation. And if a system can be exploited, it will be.

Confronting Your Inner Capitalist

The idea that wealth is about more than just having a lot of money leads us to an uncomfortable conclusion: we are all, anthropologically speaking, capitalists. That doesn't mean that we all subscribe to the political view that industries should be controlled by for-profit enterprises. But it does mean that almost all people spend their lives trying to maximize one or more forms of capital, whether resource, social, or embodied, relative to those around us. We evolved to do so—because without capital, humans are unlikely to reproduce, and without reproduction, there are less likely to be similar people in future generations.

This predilection for capital maximization is why invisible rivalry is such a nefarious quality. The fact that capital in human societies is not confined to the observable world allows us to pretend to sacrifice capital—to pretend to be altruistic—while gaining capital in another form. It also makes it easy to deny another person's altruism, for example by saying that so-and-so gave money away only for reputational benefits. The fact that we can hide our intentions is a weapon and a weakness.

To take an extreme example: a multi-billionaire like Jeff Bezos, Elon Musk, or Mark Zuckerberg has virtually unlimited resource capital, and can use it to influence technological advancement, media narratives, government policy, and even culture (digital, in these modern examples). Strongman leaders like Russia's Vladimir Putin or China's Xi Jinping may or may not have resources of their own, but they have virtually unlimited social capital within their countries, and can use it to influence or define any aspect of culture they like.[41]

Both are forms of social power. And through social power, they translate into other forms of capital and often carry reproductive benefits, tying them to traditional Darwinian reasoning.

Embodied capital can also drive resource and relational wealth growth, and in turn, power. Psychopaths arguably have high embodied capital: many of them are effective at deceiving and manipulating others, using them as means toward whatever their short-term goals are. Whether that's moving up the corporate ladder or moving themselves and their acolytes into a midwestern compound, they rely on their guile and superficial charm to maximize their capital.[42]

All of us behave in analogous ways to some degree. We spend huge amounts of time and money on education, which brings both embodied capital (knowledge and acculturation) and social capital (social connections and a diploma). We promote ourselves on social media to increase social and, sometimes, resource capital. We work to make more money or to win prestige from others. And we navigate difficult and sometimes toxic environments to place ourselves, and those closest to us, in positions of political, corporate, or social power.

We are political animals, to borrow from Aristotle, and

evolved to maximize more, and more complex, forms of capital than in any other known organism—and there's nothing wrong with that. But, as with any organism, when we take too much, and use our won or inherited capital to oppress or exploit others, we run the risk of damaging the societies we live in, much as a cancer that takes too many resources kills the body that houses it. This threat is especially great given the constant threat of invisible rivalry: hiding or ignoring our evolved drive to maximize capital allows some of the most selfish among us to compete the most successfully.

Interestingly, though, this realization can help us confront that aspect of our evolutionary story, and to solve the frustrating problem of altruism. Some people believe that humans evolved to be altruistic, at least toward members of families and our social groups, because groups of altruistic and cooperative people out-competed groups of selfish people over evolutionary time. Some skeptics believe another extreme: that altruism can't exist because, whenever we help someone, we're deceiving ourselves about our intentions. We evolved self-deceit to convince others about our altruistic motivations—and this leads others to treat us better.[43]

As with any human quality, though, we need to analyze altruism through the lenses of psychology, biology, and how people behave. Some people may have ulterior motives, conscious or otherwise, for helping another person. And some may be a part of social groups or cultures that champion kindness above any other quality—with the consequence that failing to give costs social capital.

The reality, for most people, is that they have multiple motivations when helping another person, making a donation, or choosing a life of nursing or teaching over one that brings a lot of wealth. Altruism is rare, but neither ubiquitous nor impossible. Given its combined economic and psychological elements, it just requires that we have a demonstrable motivation to sacrifice capital, of whatever kind, without receiving another form in return.

This explains why the scene in *Curb Your Enthusiasm* about donations was so memorable (at least to me). Larry was unhappy for looking like he'd made a gift for self-interested reasons, given that Ted Danson's gift was (not quite) anonymous. Both, however, were at least partly about the exchange of resource for social capital. Danson was just better at giving the appearance otherwise. But we can't collectively give social capital to people who hide their identity, who are truly anonymous givers. That's a real sacrifice, and having to resist the impulse to tell people about your good deeds is what makes that sacrifice so hard.

Insofar as we recognize the capital-maximizing nature of being a living animal, we can understand that sacrificing capital for someone else's benefit is true altruism. It's difficult to make sacrifices because it should be difficult, and it's because trading one form of capital for another is so common that so many people are skeptical that altruism exists at all. It is real, but it's difficult to talk about because it's so easy to doubt in others. Look to the effects of giving, not the source, to find it.

Shibboleth

To combat the fundamental individual drive to increase capital across the natural world, organisms have evolved elaborate mechanisms to keep exploitation in check. In many cases, these are biological functions: how an organism, like a human body, evolved defenses against selfish cells—cancers.[44]

Complex organisms such as mammals must continually fight off these rogue cells that appear, every so often, through DNA copy error during cell replication. To aid our immune systems, we have a host of strategies and resources, such as two copies of the gene TP53, which some people call the "guardian of the genome" (elephants have twenty copies, which is why they rarely get cancer). We also have immune cells that attack invaders, and a cellular function called programmed cell death that serves as a failsafe when cellular mutation occurs. Together, these functions help us to stave off most mutated cells that appear in our bodies, and it took millions of years of evolution for natural selection to refine them to near perfection.[45]

We see analogous mechanisms all over the natural world. Pine trees, for example, have evolved chemicals such as oleoresin that help to defend them against predating insects, including some species of beetle. They even communicate using these chemicals when in distress—primarily to other trees they are genetically related to—giving their relatives time to build up further chemical defenses in anticipation of an attack.[46]

These complex defensive strategies have worked to protect them historically, except now, in Canada and other places, invasive pine beetles have evolved a countermeasure that allows them to overcome the chemical barriers. Coupled with climate change, this has led to an explosion of predation of pine trees in North America, and global warming is allowing the pine beetles to spread further than before.[47]

This is an example of what is called a coevolutionary arms race. These resemble a real arms race—as with the United States and the Soviet Union stockpiling nuclear weapons during the cold war—except they occur over evolutionary time, often between two species, or even between two types of individuals within a species.

In some species, such as the centrarchid sunfish, large, dominant males tend to monopolize access to females by defending a territory. They let females into these territories to maximize their mating opportunities, and fend off smaller males who attempt entry. It's a good strategy, assuming you are big and aggressive enough to defend a space. But some males who aren't so big have evolved a workaround: they started to resemble females. When these sneaky males approach the territory, the large male doesn't take notice—giving the smaller one access to the female, too.[48] This creates a coevolutionary arms race between the female mimics and the territorial large males: the former needs to resemble females enough to get through, and the latter needs to recognize them without expelling a real female.

The contest between deception and recognition is common in nature. Cuckoos, for example, lay their eggs in the nests of birds of other species. For any non-cuckoo, the obvious thing to do to avoid wasting precious resource and energy is to get rid of the alien egg. Evolution has made things difficult, though: cuckoo eggs and chicks can resemble their hosts, sometimes even mimicking the birdsong of the host's own chicks. This leads the oblivious host to foster the invader—a brood parasite—sometimes at the cost of losing its children, if the invading cuckoo pushes the other eggs out of the nest. Again, the consequence is an arms race: some hosts have learned to recognize and push out alien eggs, or to attack parasitic birds—or fish, or insects, or any class of organism with brood parasites—when they see them.[49]

This pressures the brood parasites to look more and more like specific host species—with the risk that, if they prevent any host from reproducing entirely, driving them to extinction, they won't have anyone to predate on. With selection, balance is everything: and as we are starting to learn, over-exploitation of any resource can damage our future.

Humans, too, have mechanisms for preventing interlopers from entering our social groups and taking our resources. There's an extreme example from the Bible: the story of Shib-

Common cuckoo eggs, which resemble host eggs (Tomas Grim)

boleth. In this legend, one tribe defeats another in a battle, and then guarded the passages of Jordan. When someone approached the passages, the victors asked them to pronounce the word "Shibboleth"—knowing that only members of their own group could enunciate the "sh" sound. This may sound implausible, but the method was at least purportedly effective. Of members of the losing tribe, the Ephraimites, the book of Judges says: "and he said Sibboleth: for he could not frame to pronounce it right. Then they took him, and slew him at the passages of Jordan: and there fell at that time of the Ephraimites forty and two thousand."

Reverberations of this strategy for sorting group members live on in modern society—to the point that "shibboleth" refers to a custom or behavior that demarcates cultural group membership. And although we may not slay anyone for incorrectly pronouncing an unvoiced fricative (such as the "sh" sound), we do use phonetic measures for determining where people are from. Some immigration departments in Europe, including the United Kingdom, use the Language Assessment for Determination of Origin (LADO) when making decisions around some asylum applications. If someone applies for asylum while fleeing a war-torn country like Syria, they may be subjected to this test, through which linguists analyze the applicant's speech patterns to determine their true country of origin. This may be overtly unlike the shibboleth story, which leads to direct killing, but the outcome of a LADO evaluation may result in someone being deported to an unsafe country.[50]

Interestingly, some animals have similar strategies for preventing out-group members from entering their social groups. A study by Alison Barker and colleagues showed that even naked mole rats are (in human terms) xenophobic, and show hostility toward rats who don't chirp in their own colony's "dialect," which these animals learn from a young age. This is essential for mole rat colony survival: the shortage of available resources, coupled with limited reproductive opportunities, translates to great hostility to interlopers who don't share a high proportion of genes with the members.[51]

The Cultural Immune System

More generally, human cultures appear to have developed analogous mechanisms for stifling deceptive or exploitative strategies for maximizing capital, similar to mechanisms found among other animal species and even on the level of cells. This is perhaps one reason that many modern researchers are interested in comparing these biological and cultural strategies. In the case of people, however, breaking rules or norms for personal benefit is more complicated than just a cell taking more than its share of resources from a body. We can take more than our fair share from a hunt, yes, but we can also betray our social contracts in other ways, such as evading taxes in Western societies or being cowardly in war among hunter-gatherers. The difference in larger, stratified societies, however, is that invisible rivals can more easily fade into a crowd: one problem with anonymity is that it makes it more difficult to detect cheaters, and consequently gives people an opportunity to cheat.

In smaller-scale societies or cultural groups—the circumstances in which we have spent nearly all of our evolutionary history—norms are often easier to enforce. Adultery, for example, is a serious offense in many cultures. The biological explanation for this is obvious: in many places, males want to know that their wife's child is their own, otherwise they risk wasting investment in another person's child—much like a warbler might waste resources investing capital in a cuckoo's offspring.[52]

In some groups, people have contrived complex and unpleasant strategies for maximizing what's called paternity certainty, or the impression, justified or otherwise, that a father's child is his own. Among the Dogon people of West Africa, women construct *punulu*—"menstruation houses"—outside their villages, which menstruating women live in for the duration of their periods. The ritual relates to the perception that menstruating women are unclean, but in practice, it allows husbands and their families to ensure no other men have access to their wives before they become pregnant.[53]

Orthodox Judaism has a similar practice that relates to the perception that menstruation is unclean: the *mikvah*. This is a pool of spring water that women must visit after their periods, where they are declared clean after immersing themselves. Again, whatever the symbolic importance of the practice, it has evolutionary consequences. If a husband knows when a woman menstruates, he can better gauge her fidelity when she becomes pregnant.[54] And generally, the amount of capital a father invests in his offspring across cultures does seem to depend on facial similarity, as well as odor, which are two mechanisms by which people are thought to be able to detect kinship.

Norms around marriage and mating also tend to conform to local socioecological surroundings. In some polyandrous societies (ones where a woman has multiple husbands or partners), such as the Tibetan-speaking Nyinba people of Nepal and other groups in Tibet, it's common for two brothers to be married to one woman. Jealousy over the woman is thought to be low. In these cases, it still makes sense for both brothers to help raise any child the woman has, because if one brother is the father, the uncle still has a 25 percent genetic relationship with the child. In the case of multiple children, being a father or uncle to any or all diffuses the incentive to invest between the brothers.

In some cases, as with speakers of the Carib, Pano, Tupi, and Macro-Je languages in South America, brotherhood among a woman's sexual partners isn't so common. Whereas in many Western cultures that's likely to lead to violent jealousy, these groups of people believe in what's called "partible paternity"— or the idea that all men who have sex with a woman around the time of conception are the child's father. Paternity is, for these groups, a quality that multiple people can have.[55]

Again, at the surface, this doesn't make much sense from an evolutionary point of view. But then again, the genetics of it can work: if you are one of four partners a woman has, and that woman has four children, there's a good chance that at least one is yours—and that each of the four fathers has a duty to protect and invest capital in your child. This is a particularly valuable relationship in groups where warfare is common. A child with multiple fathers is unlikely to be left fatherless if any one of them dies fighting, benefiting the mother, the children, and the fathers' genetic legacy.

There are historical examples of partible paternity in ancient Western cultures, too. In the *Gallic War*, Julius Caesar wrote: "All the Britons, indeed, dye themselves with wood [woad], which occasions a bluish color, and thereby have a more terrible appearance in fight. They wear their hair long, and have every part of their body shaved except their head and upper lip. Ten and even twelve have wives common to them, and particularly brothers among brothers, and parents among their children; but if there be any issue by these wives, they are reputed to be the children of those by whom respectively each was first espoused when a virgin."[56]

Polyandry remains the rarest mating system across human cultures, with some estimates placing the figure around four per five hundred societies, though there is some dispute in the anthropological literature.[57] It's also a practice that is linked with inhospitable environments. If a woman owns a large piece of land that can provide resources for a family where dividing the land up cannot, it makes more sense to avoid inheritance to sons or brothers. Polyandry is just one of the diverse stories of human history where our cultural adaptations help us to thrive when other practices have not, for ecological reasons, worked.

Punishment is another of these adaptations, and one moreover that is found worldwide. Given the centrality of marriage and mating norms to any person's genetic legacy, breaking these rules can be dangerous. A study by Zachary Garfield and colleagues found that among 131 societies across the world, violations of adultery norms occurred in about 60 percent—

and the most common punishments were physical attack and execution.[58]

It makes sense, then, that norms around marriage and fidelity help to maintain cohesion in societies. The anthropologist Joseph Henrich argues that this is why monogamy is so common: it helps to ensure that everyone has a good chance of securing a partner, reducing the risk of violence unmarried men can pose. And these risks can be great, as the modern movement of involuntary celibacy ("incel") shows: the perceived lack of access to potential mates can create a lot of anger among those who don't view themselves as sexually successful.

Most norms have analogous functions.[59] Whatever their other purposes, taboos, dietary restrictions, religious rites, and warfare practices help to regulate behaviors and keep society functioning. Yet the forms punishments for norm-breaking take accord, unsurprisingly, with the forms of capital found in specific societies. We punish by inflicting physical damage (harming embodied capital), forcing people to make payments to others or institutions like religious bodies (reducing resource capital), and by hurting a person's reputation (social capital). Execution ("capital punishment"), while obviously physical, is arguably all three.

Societies can therefore be said to have cultural immune systems. Our diverse histories and environments we live in drive the forms that our norms take, which are probably the consequences of strategies that exploiters have used previously in a given environment. These practices help to protect us from ourselves and each other much as the body's immune system exists to eliminate hostile germs and cancers.

The crossover of biological and cultural adaptations that help to thwart disease illustrates this. The feeling of disgust evolved to help us avoid substances that can make us sick, and we also have group-level behaviors that help to prevent the spread of sickness. Quarantine, for example, is a mechanism that helps to stymie disease transmission but which requires cultural acceptance—or the use of force—for successful implementation. The whole field of public health, in this light, is arguably a set of cultural adaptations for helping us to collectively reduce disease burden in our populations. And historically, these measures have been essential: infectious disease has, according to medical anthropologists, probably killed more people than all wars and natural disasters in history put together.[60]

Cultural adaptations that stymie individual selfishness work in a parallel way. We evolved social norms to help our groups function more cohesively just as we evolved norms like quarantine. Punishment is one way we maintain those norms—and the threat of punishment can be an effective deterrent that maintains obedience.

This is true across societies and across history. Norms of sharing resources among hunter-gatherers help to reduce resource inequality and maximize everyone's chances of survival, even if better hunters have more reproductive opportunities. Ancient Mesopotamian groups like the Babylonians had written legal systems—the Code of Hammurabi, for instance, which dates to 1755 BCE, prescribes equality in crime and punishment (and is where we get the phrase "an eye for an eye" from). And today, people living in large, industrialized economies pay taxes in part for public goods that (at least in theory) can help to reduce resource capital inequality.

The Golden Rule

Maori beliefs around gift-giving are an example of how we take cultural immunity past simple mechanisms of enforcement like punishment. Growing up in a culture leads many of us to believe in the rules we follow as part of a larger understanding of the world. And again, these beliefs cross over with the hidden world of disease: not reciprocating a gift, the Maori believe, can lead to illness.

Similar norms promoting reciprocity worldwide, such as the Golden Rule, help to instill beliefs in us about how best to treat others—and moreover that treating others well is likely, over time, to benefit us, too. In the academic literature, this is called "internalization," and occurs when cultural norms become more than just things we do. Norms help to determine what we believe to be right, and what sort of people we are.

Religion is a famous and universal example in ethnography. We grow up Jewish, Islamic, Christian, Buddhist, Manichaean, Zoroastrian, Sikh, or any of the other estimated four thousand religions in the world. And during this upbringing, we're taught not only what to do, but why it matters: we're taught to be ethical, not just obedient. For Aristotle, this was a key distinction. It mattered, for his ethics, not only that you understood what rules to follow, but why it's important to follow them. And a close study of the evolution of social norms helps to make this picture clear. Norms are the social adhesives we depend on. They help people to stop the endless pursuit of capital and to focus on what's best, for everyone, in the longer term.

How these norms evolve is, however, an essential part of the benefits they offer. Not all norms are good. People rely on norms to justify rape in warfare, violence in blood feuds, and even tax evasion among the hyper wealthy—it's just about what is normal for a given group of people. One view in economics is that people across society struggle to create the rules around capital that benefit them most. This often results in class struggles: the wealthier upper classes aim to design rules to benefit themselves at a cost to the poor; the poorer working classes aim to create greater equality.

Manvir Singh calls this "subjective cultural evolution." We design the rules that benefit us most, and when those come into conflict with the rules other, competing people want to institute, we see either a victory for one side or a synthesis of the rules.[61] The Golden Rule is the oldest, most common, and most fundamental of these. Older, widespread norms are likely to be the products of thousands and thousands of years of self-interested rule design. They are the signals from our shared past about what has survived, and what is likely to work best to maintain a functioning society.

But like every human institution, norms work well to stop people endlessly maximizing capital only if there is a motivation to follow them. Just as with biological immunity, the cultural immune system has holes that allow for deceptive exploitation—that allow for invisible rivalry. We still get cancer despite millions of years of genetic evolution; the temptation to cheat in social interactions is, likewise, omnipresent. This is evidenced even by the gift-giving beliefs among the Maori: because gifts must be repaid, people sometimes ruin other families by giving them things so valuable that they cannot be reciprocated.

The same kind of arms race that led to our biological immune systems also led to our cultural ones—but as at the biological level, the cultural race is ongoing. Invisible rivalry is the product of cultural evolution over hundreds of thousands, if not millions, of years. Every tactic we develop to deter, find, or punish exploiters is met by an equal or better strategy to remain hidden, precisely because acculturated people are specialists in their own social norms, and so are best placed to develop new modes of exploitation.

Capital, of whatever kind, motivates our behaviors, and we are all, at our core, capitalists. Yet because invisible rivalry permits the effective exploitation of large, stratified societies, reliance on social norms and punishment is unlikely to help us in the future. This is because, given our evolutionary story, invisible rivals will always find a way to hide among us—and follow rules only for as long as they benefit from doing so.

5
The Power of Darkness

A common theme in literature, film, and television is the corruption of good people. Or maybe more simply: the corruption of goodness—a concept of pure morality we find in some of the world's major religions. The snake tempts Eve; Eve tempts Adam; both are cast out of paradise.

Lady Macbeth serves a similar purpose in Shakespeare's famous play: she drives the title character to commit progressively worse moral transgressions, setting him up for his inevitable demise. The Russian writer Nikolai Leskov bases one of his major works, *Lady Macbeth of the Mtsensk District*, on Shakespeare's work, and shows the central character, Katerina, making progressively more vicious decisions, and influencing others around her—out of what seems to be boredom—to do the same.

We see this theme in recent productions, too. AMC's famous TV show *Breaking Bad* shows Walter White's descent from goodness (or weakness?) into murderous criminality. And in 2022, the filmmaker Florian Zeller wrote the short-running play *The Forest*, which explored how a single poor decision led an otherwise harmless central character into an inescapable maelstrom of corruption and violence.

Maybe darkest of all is one of Leo Tolstoy's lesser-known

works—a play with the admittedly melodramatic title *The Power of Darkness*. In this play, initially banned in Russia, Tolstoy shows how an ordinary man is tempted, through both opportunity and by a Lady Macbeth–like character, to commit successively worse actions—culminating in a graphic description of infanticide.

The play inspired a more subtle take by the late Nobel laureate Isaac Bashevis Singer, who wrote a short story with the same name. Many of Singer's short and longer stories explore the theme of temptation, real and mythical. We are plagued by demons, hormonal or imagined, that lead us to disregard our moral inclinations and act selfishly, often at the expense of others. And once we give in, it's hard to turn back.

These portrayals, among many other good features, analyze, as much as is possible, the internal states of people who make all the wrong choices. We have a morbid fascination with these characters because we can relate to them. It's not hard, when opportunity permits it, to cheat on a test, steal a stick of gum, or use a large language model to write a difficult essay. Most people who say they don't feel temptations are either lying or deceiving themselves—it's just part of how we evolved to try to take a little bit more for ourselves, especially when no one is watching.

These stories help us to understand how to relate to psychopathic, sociopathic, or narcissistic people who commit crimes in modern society, too. In the late 1990s, the *New Republic* fired an increasingly famous journalist, Stephen Glass. Until that point, he'd been a rising star in the writing world: he'd graduated from an Ivy League university where he was executive editor of the student newspaper. He then worked in several prestigious journalistic roles before taking a permanent job at the *New Republic*, where he wrote several high-profile and, later, controversial feature stories.

The features he wrote in his last position as a journalist led to objections from many of the institutions he criticized. Questions emerged about the reliability of his sources. And eventually, after a feature he published in 1998 about the world of hacking, his own bosses realized there was something wrong. Glass had invented a teenage hacker who didn't exist, and then created a website for a fake company that, he wrote, had hired the hacker. Charles Lane, Glass's editor at the magazine, noticed inconsistencies in his work—and found out that a supposed source for the feature was Glass's brother.[1]

This was merely an inverted smoking gun, but his career, it turned out, had been built on fabrications. Of forty-one stories in the *New Republic* with his byline, two-thirds were at least "partially fabricated," according to the *Washington Post*.[2] These included a made-up "First Church of George Herbert Walker Christ," which was allegedly a cult obsessed with George H. W. Bush, the former U.S. president, as well as non-existent sources for an article that attacked D.A.R.E., a well-known but controversial anti-drug program in the United States.

After Glass had been fired, Lane told a reporter: "We extended normal human trust to someone who basically lacked a conscience. . . . We thought Glass was interested in our personal lives, or our struggles with work, and we thought it was because he cared. Actually, it was all about sizing us up and searching for vulnerabilities."[3]

It's easy to point to cases like this—and there are too many to list—and say that they are just extreme examples, and the perpetrating psychopaths always pay a price. The difficulty, however, is confronting the reality that we don't know how many people cheat in small ways and are never caught. We can't, to misuse a thesis of Ludwig Wittgenstein's, talk about what we don't know.

As with acts of true altruism, anonymity, particularly in

large societies, makes deception difficult or impossible to uncover. People are, despite extreme argument otherwise, mixed: sometimes we help, and sometimes we cheat. We obsess over extreme examples like Glass's partly because we see ourselves in them. And yet the project of uncovering a "true" human nature marches on—and often, in modern work, focuses on the better sides of our nature (the late Frans de Waal's *Good Natured* is a good and informative example). To understand the sides we want to ignore, however, we need to look beneath the surface.

The Swiss Man's Money Games

The surface of much current research and popular writing creates the impression that we are kinder and more cooperative today than ever in our history. One columnist from the *New York Times* argued that people are more generous than most would guess. He cited a study from 2023 that explored how people from different countries spend money given to them unconditionally.[4] The authors gave two hundred people from three low-income countries—Indonesia, Brazil, and Kenya—and four high-income countries—Australia, Canada, the United Kingdom, and the United States—$10,000 each, and told them to spend it however they wanted. The experiment's only groupings were by country and by another variable: half were asked to record their spending on the website formerly known as Twitter with the hashtag #MysteryExperiment, while half were told to keep their spending private (this did not include, if you read the study closely, the participants' friends and family).

Across countries, people spent, on average, about half the money ($4,780) on themselves, $3,239 on family members within their household, $910 on friends, about $1,800 on donations, and $466 on strangers and acquaintances. These amounts didn't differ, again on average, between low-income and high-income countries, and people in the public display group—those asked to tweet about their spending—didn't spend differently overall, though they spent a bit less on family in their household.

The data were hard to read in the paper. Past giving high-level information like the arithmetic mean and standard deviations—basic statistical descriptions—it was hard to tell how people varied in their spending. I contacted the lead author, Ryan Dwyer, who kindly sent over a graph, shown here, that depicts the variance in which I was interested.

The graph is a histogram that charts the amount of money participants spent on people and things outside their household (the x-axis, I assume rounded to the nearest $500 increment), along with the number of participants who spent at the given increment (y-axis, "Count" on the left-hand side). The dashed line is the mean (the arithmetic average), which was $3,678 for all participants across the study.

The histogram shows that the most common result—the mode, in the language of statistics—was $0. More than 25 people out of the 200 spent nothing on anyone or anything outside of their own household, which is around twice the number of people that spent the full $10,000 outside their household. So while one statistical average, the mean, gives a heartening result, the mode, another average, is a lot less heartening—and "the average person spends nothing on people not in their household" is a less clickable, and badly written, headline. Both are averages, but one fits a desired ideological framework, and one does not.

The study's findings reflect a huge and growing body of literature that aims to answer the same questions: Are people generous? Do we conform to the predictions of Pyotr Kropotkin or Thomas Hobbes? The number of research projects in

The Power of Darkness

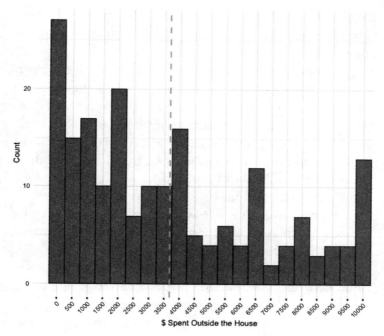

Histogram of results from a generosity trial. The x-axis indicates the amounts of money people spent, out of $10,000 given, outside of their own household; the y-axis gives the number of participants who fall into each $500 category on the x-axis. The dashed line is the median (one form of average). (Ryan Dwyer)

this area is so large that some people recently developed a website—the Cooperation Databank ("CoDa")—that allows for the real-time analysis on the known work done in this area to date. At the time of writing, that's about 1,809 academic papers representing 2,636 studies and more than 350,000 participants.[5]

The projects are varied and often clever in their design. And perhaps too often they quote Adam Smith at the beginning. "It is not from the benevolence of the butcher, the brewer, or the baker that we expect our dinner, but from their regard to their own interest" is a favorite in the literature, and betrays a target, or even bias, on the part of the researchers interested in human pro- and anti-sociality.

Some researchers use simple designs: they give one person some money and a choice: keep it, or give some or all of it away to another person. Sometimes participants play more complex games with multiple people. Sometimes they can punish each other for not behaving prosocially. All of them, however, start with the fundamental assumption that the rational thing to do, evidenced by the obsession with Adam Smith, is for a participant to keep as much money as possible.

People consistently betray this doctrine. Some act a bit prosocially, others very. Yet across the literature, at the surface, there's little evidence that anyone is consistently selfish. One analysis of more than 650 studies conducted in the United States between 1956 and 2017 revealed an interesting trend: people appeared to have become more generous over the sixty-year period. This was surprising to the researchers, given that public trust was purportedly lower, and resource inequality was demonstrably higher, in 2017 than in the 1950s.[6]

Some people, notably the anthropologist Joe Henrich, object to how these studies are conducted. Too often, participants in these projects—often called "economic games"—are from rich Western countries. Henrich invented the term WEIRD (Western, educated, industrialized, rich, and democratic) to refer in part to the lopsided participant samples used to make claims about human nature.[7]

He and his colleagues argue that we need to conduct economic games in other places instead—places where people live in environments that more closely resemble the circumstances humans evolved in. If *Homo sapiens* emerged about 300,000 years ago, the past 150 years of industrialization are only a tiny speck in our history.

To this end, anthropologists have conducted economics games experiments in groups of people from across the world. A major study in 2001 of fifteen small-scale societies—including hunter-gatherer groups like the Hadza and Aché—found that selfishness, as measured by prosociality in the games, was rare. People across groups showed a willingness to share the resources given them by the researchers with others in several different economic games.[8] There were some differences among the groups, but none were overtly, on average, selfish. This and the similar studies that followed seemed to confirm the idea that prosociality is common to being human, though the reasons for that prosociality are likely to depend on cultural context.

One researcher involved in the 2001 study, Ernest Fehr, who has spent his academic career at the University of Zurich in Switzerland, has conducted numerous economics experiments with populations across the world. Fehr has used his findings to justify his belief in what he calls "inequity aversion"—or the preference, ostensibly common to all people, for a fair distribution of resources.

Yet the participants' impressions of what these experimenters from rich countries are up to is not always consistent with this interpretation. Polly Wiessner, who works with the Ju/hoansi/!Kung Bushmen of the Kalahari, wrote in 2009 the following: "I decided to play experimental games with the Ju/hoansi after discussions with Ernst Fehr of the University of Zurich. To protect myself from the many requests that arise when resources are available, I explained the project to the Ju/hoansi by saying that there was a man in Switzerland who wanted to see how they played experimental games. I made it clear that it was his interest, that I did not care at all what their decisions were in the games, and that there would be no consequences for how they played. Because so many odd research projects have been carried out among the Ju/hoansi, this did not seem strange to them at all, and they were all very keen to play. To this day they ask when is the Swiss man going to send more 'money games.'"9

Wiessner, who conducted several economic games with the Ju/hoansi, also tracked how they spent the money they were given after the experiments were done. She found differences in how people behaved in the anonymous games and how they behaved once they had money. They were stingier when playing the games, but more generous when they'd won their earnings. Notably, Wiessner writes about the participants' interests in the reputational elements of the games: "A few asked me once more if it was *really* true that their identity would not be revealed; with confirmation, they slid more coins, one by one, over to their own sides. Occasionally the subject would hesitate and say, 'Are you sure you are not deceiving me?'"

Wiessner argues that the behaviors in the "game" and "real life" were different because the games removed the cultural context in which giving normally takes place. Divorced from their guiding cultural tenets, it was easier to be selfish; when spending money they'd won, the cultural requirements of generosity were forceful. All many people really cared about, when Wiessner spoke to them, was whether how they behaved increased the likelihood that they would be invited to play more money games.

Under Western Eyes

The cultural contextuality of prosociality—as observed across these studies, when you look more closely than just the surface-level observation that people do not adhere to cold, economically rational selfishness—makes clear our collective and intuitive inability to divorce resource from social capital. We aren't concerned only, as the overused quotation from Adam Smith implies, with taking as much money as we can when it's available. We also want to send messages to others, particularly when we think we are being watched or studied.

Researchers are aware of this, and have attempted to account for it when designing games, though it's arguably impossible to remove the feeling you are being watched once you sign up to be a participant in an experiment. Still, the clever designs of many of these projects tell us a lot about when people are likely or less likely to cooperate in economic games.

We can, as the 2001 study shows, find cultural differences in how people play, even though we are removing people from the social contexts in which they normally interact. If experimenters are watching everyone, then being watched can't explain why people from one hunter-gatherer group are more generous in an experimental setting than are others. Something else, like local social norms, is needed to explain the findings—even if those norms are weaker influences than they would normally be in a non-study context.

This was true of a set of studies conducted in the 1980s by the late sociologist Toshio Yamagishi.[10] These experiments, which have been highly influential on economic games ever since, suggested that people living in states with more surveillance are less likely to show trust in lab settings. Specifically, Yamagishi was interested in how much trust and cooperation American and Japanese participants showed in two experimental settings representing a real-life social dilemma, like whether to contribute to a public goods project, such as a community park. In the first setting, participants were penalized ("sanctioned") for not cooperating, which involved contributing to the public good; in the second, there was no sanction.

Americans—who, at least at the time, did not live in a society with a high degree of surveillance and monitoring—were more trusting, and more cooperative, when the sanctioning system was removed. Japanese participants, however, were less cooperative and less trusting without the sanction. This accords well with predictions from the problem of opportunity, but goes a step further: in a society characterized by a high degree of surveillance, we can count on people to cooperate only insofar as they are worried about being observed or punished for not cooperating. Without surveillance, people rely more on local norms to guide their trust and prosociality. Yet with the rise of facial recognition technology and more general concerns around surveillance capitalism—many people feel they are being watched by the technological tools like smartphones they rely on—it would be interesting to see if Yamagishi's findings remained accurate in contemporary American society.

Other creative research shows darker trends than a surface-level analysis of cooperation studies might suggest. One study showed, for example, that some people stop behaving prosocially when they believe their fellow game players won't know about their choices.[11] The researchers showed this in an economic game called a "dictator game," which has two players. The first player—the dictator—is given an amount of money, usually something like ten dollars, by the researchers, and can choose how much to give to the second player, who has no choice but to accept whatever (if anything) is given.

Researchers interested in the evolution of cooperation often point to dictator games as proof of the human preference for fairness: dictators often give money even though there is no drawback—within the game's structure—for doing so. This study, however, gave dictators another option: they could pay one dollar of their allotted ten to leave the game without the second player's knowledge, keeping the remaining nine for themselves. About one-third of dictators chose this option, even though they received one dollar less than if they left the game with the full ten—paying the price, instead, of the second player knowing they'd done so.

The researchers then created a "private" game situation where the second player didn't know they were playing a dictator game at all, though the dictator did. In this setting, not one dictator chose the "quiet exit" option: they didn't need to pay for anonymity.[12]

Another study design is reminiscent of F. Scott Fitzgerald's *The Great Gatsby*—specifically, Dr. Eckleburg's faded eyes on a billboard, and the fear of divine omniscience that they induce.[13] In this experiment from 2005, participants sat in a cubicle and played an economic game on a computer: they could choose to donate, or not, to a simulated public good. (As with almost all of these games, the rational decision—economically speaking—is not to donate, but regardless to capitalize on the public good, which is usually extra money.)

There's a twist, though: for some participants, the researchers placed a pair of watchful eyes in the cubicle, and for others they did not. They found that people in the cubicles with the eyes donated more to the public good than did others.[14] The feeling of being watched—in addition to already being part of an experiment—contributed to their prosociality.

There are far too many studies, with too many clever designs, to discuss even a hundredth of them here. Yet the studies that dig deeper than just asking whether people behave in accordance with a quotation from Adam Smith show there's more to the picture than inequity inversion, a preference for fairness, or any other academic way of saying that people are inherently generous. Few people would go to Central or Regent's Park on a sunny day and assume that the happy faces they saw represented the average moods in New York or London. The happy surface is misleading, and we need to go deeper still to find out what drives human behavior, under experimenters' eyes or not.

The Rationalizing Man

A more convincing thesis than inequity aversion, at least to me, is that fairness is a yardstick, not a preference. We don't have an innate preference for equality, but rather we measure our own and other people's behaviors against what we believe is fair. Viewed in this way, perceived fairness is a barrier to exploitation that invisible rivals seek to find a way around, not an end that all people desire.

This extends the common and old description of humans: the "rational man," which relates to the concept of *Homo economicus*. As Adam Smith and many other economists predict, the "rational man" ideal suggests that people evolved to behave in a calculated way that benefits them. The thousands of economics games experiments conducted to date suggest this isn't correct, though they don't suggest a complete commitment to prosociality, either.

We are, at least when it comes to fairness, rationalizing, not just rational. We make complex judgments about what we think people, including ourselves, deserve in different situations, taking into account more than only our selfish or prosocial desires. The effect known as "moral licensing" (or credentialing), which researchers have observed in several studies, helps to illustrate this.[15]

A study in 2001 showed, for example, that giving a person the opportunity to express a moral view can increase the odds that the same person behaves in a biased way.[16] In the experiment, which evaluated sexism in hiring practices, one group of men had the chance to openly disagree with a sexist statement; a second group wasn't given this chance. Yet those in the first group were more likely to select male candidates for a stereotypically male position than were those in the second group. Airing views on egalitarianism led to biases in action against it.

Another study suggested that even imagining making a donation to a charity can increase the chances that a participant makes a luxury goods purchase.[17] And a third found that people are more likely to cheat on a math test if they can rationalize about why cheating is acceptable.[18]

It seems that creating the impression, to yourself and others, that you believe in fairness gives you the cognitive tools necessary for acting unfairly. To the degree that we can rationalize our actions—defend them in an imaginary court, as it were—we can override our sense of fairness.

This yardstick view of fairness accords with some notable findings in variations of economics games, too. In dyadic games where the outcome relies on how both players behave, rather than just one as in the dictator game, people accept divisions of money that reflect perceived ownership of the potential earnings.[19] Even other primates react to unfairness in experiments: brown capuchin monkeys will turn down unequal offerings of food when they've done as much on a particular task as those given more.[20]

Rationalization is the tool we use to circumvent or distort fairness for our own benefit, and it sometimes even blurs our understanding of what counts as fair. "Equal pay for equal work," a common aphorism, also appears in the Treaty on the Functioning of the European Union. Interrogating it a little, however, can render it difficult to implement: What if one worker needs to live in a more expensive city? What if one has three children to look after and another is single? What if there is huge variation in the age and experience of employees?

None of these questions undermine the force of the rule,

but they do force decisions about how the rule is applied.[21] This problem is more than theoretical. In contemporary politics in the United Kingdom and the United States, perceptions around fairness drive claims on divisive topics like taxation. Members of the Conservative party in the United Kingdom argue that it's only fair to raise taxes on everyone equally. And in 2022, after the failed tenure of the former prime minister Liz Truss, the chancellor of the exchequer raised national insurance taxes, increasing payments from most working people in a numerically equal manner.[22]

Some moderates and nearly all members of the left-wing Labour party argued that this increase was not fair: People who make less should not have to face the same tax increase as people who make more. Can we appeal to an idealized concept of fairness to say who is right? Or is the concept of fairness being used, in one or both cases, to further a political agenda toward or away from equality?

Political and market forces in modern society consistently distort our conceptions of fairness, leaving us at risk, through laziness, ignorance, or indifference, of exploiting others—and of being exploited. In 2013, two researchers showed the damning effects of market forces explicitly in a study of how these forces can distort ethical decision-making.[23] They used real mice known as "surplus mice," which are bred for experiments like studies on genetics, though for whatever reason they can't be used as intended. (The authors note, I think as an ethical defense of conducting this study at all, that these surplus mice are killed by default; why that is relevant will be clear in a moment.)

The study had three groups: an individual group, a bilateral market group, and a multilateral market group. In each of these, the researchers explained to participants that they were choosing whether a mouse lives or dies—though they did not tell them about what normally happens to surplus mice until after the experiment. They did, however, show participants a graphic video of how laboratory mice are killed; they also told each that the mouse's life was entrusted to their care. In the individual group, they were then given a choice: receive no money and the mouse lives, or receive ten euros and the mouse dies. Of 124 participants in this group, 45.9 percent chose to receive the ten euros.

This already isn't a promising figure, but introducing a market element had an effect, too. Participants in the bilateral market setting were split into groups of two: one buyer and one seller. They were collectively given twenty euros and haggled for as long as they liked over a price for the mouse. If they agreed, the seller received the sale price and the buyer received the remaining money; the mouse was then killed. In this setting, 72.2 percent of sellers accepted a sale price of ten euros or less. The multimarket setting was identical to this, except it had more buyers and more sellers; the findings, however, were similar: 75.9 percent of sellers were willing to let a mouse die for ten euros or less.

The authors suggested that market forces explain the difference in how individuals behaved compared with those in pairs or groups. An economic setting involves multiple people and may spread the guilt of one person's decision: two or more people become complicit in the mouse's death. The way markets work may also create the impression that otherwise unethical behaviors are acceptable in a particular society. If everyone is trading in mouse lives, it seems less wrong to do so. And finally, the interactions inherent to a market—bargaining, arguing, socializing, and so forth—may become the focus of buyers and sellers, rather than the fact that an animal's life depends on them.[24] Market forces, otherwise put, can lure us to behave in ways we'd normally call unethical.

A real-world example is cocaine, which is widely enjoyed as a recreational drug in countries like the United States and the United Kingdom. Yet the *Guardian* reported in 2005 that every gram of cocaine snorted is linked with one murder in Colombia, which historically has illegally supplied much of the Western world's habit.[25] The murders are not limited to people involved in trafficking, either: parents, children, and even babies are among the victims.

Our ethics are therefore different—or maybe, our morals manifest differently—depending on the setting. It's easy to say that buying palm oil or buying clothes made in sweatshops is wrong; it's something else to avoid doing both entirely. We depend on markets for our comfort and livelihoods, and ignoring the less pleasant sides of what markets mean for our ethics is part of living in the twenty-first century while maintaining our sanity.

This is a frustrating truth most people already know about, and it connects to our impressive ability to rationalize away the unfairness by which we are surrounded. The discussion turns dark quickly, though. Ernst Fehr, the Swiss man whose games the !Kung love playing, works for a publicly funded university. Switzerland, however, which was ostensibly neutral in World War II, took ownership of an unknown sum—thought today to be in the many billions of dollars—from Jewish people murdered by Nazi Germany in the Holocaust. And although Switzerland agreed to pay over a billion dollars to the families of these victims in 1998, the full extent of Swiss seizures of these victims remains unknown, with information on at least two million bank accounts missing.

In 2004, U.S. judge Edward R. Korman wrote a startling opinion concerning the case against Swiss banks, over which he presided for several years.[26] "The 'Big Lie,'" he wrote, "for the Swiss banks is that during the Nazi era and its wake, the banks never engaged in substantial wrongdoing." He noted that, following the landmark agreement in 1998, Swiss banks continued to dodge repayments, lie about missing documents, and consistently misrepresent the funds they had appropriated or transferred to the German Reichsbank during the war.

Korman highlighted how Swiss policy allowed banks, like the now-defunct Credit Suisse, to avoid account access and destroy evidence of accounts, allowing them to appropriate whatever funds were in them. Where relatives of Nazi victims sought information about accounts, banks charged increasingly high "search fees" over the latter half of the twentieth century—from 25 francs in the 1950s to 750 francs in the 1980s. "It is important to reiterate that the Swiss banks' devotion to secrecy and their repeated acts of stonewalling were not based on principles— they were profit-driven," Korman wrote.

Yet, ironically, a website run by a digital marketing agency that advertises Switzerland as an educational destination points to the financial services industry when discussing the country's historic economic success. "Another post-war change in the Swiss banking system was the formation of the legal secrecy policies of the banks, which made Swiss banks a favorite destination for many wealthy Europeans, including European Jews fleeing the Nazi persecution," the website states. "Today, the finance sector remains an important sector of the Swiss economy generating more than 10% of the country's GDP [gross domestic product]."[27]

It is interesting, then, that the money games run by academics at an important Swiss university are said by some to show an innate human preference for equality: an absurdity, perhaps, that only a rationalizing man, steeped in a market economy, could adopt.

Around the Yardstick

All institutions created by humans have flaws that permit their exploitation. Some months ago, a friend told me she believes religions—to take an emotive example—are like cults. Major organized religions, to her point, have almost too many scandals to enumerate, from the historic use of religious titles to deceive and exploit followers to more modern cases, like pedophilia cover-ups in Catholicism. You would be hard-pressed to find any religion, however attractive in ideology, that didn't have examples of its leaders using doctrine to enrich themselves, exploit their followers, or suppress resistance to their leadership.

Yet unlike cults, which operate around a central psychopathic character, religion—or more broadly, religiosity—is a powerful unifier that can, at least when it works well, help people to overcome their selfish instincts. It's probably for this reason that Jonathan Swift, the seventeenth- and eighteenth-century Anglo-Irish clergyman and famed satirist, wrote that "Religion being the best of Things, its Corruptions are likely to be the worst."[28]

Those corruptions have, over the past several centuries, led many people to claim that organized religion exists entirely to suppress and exploit already marginalized groups of people. Movements like Humanists U.K., the American Humanist Association, and New Atheism repeatedly attack the continuing influence of religious bodies in Western democracies. Humanists U.K., for example, argues that religious doctrine should not be on school curricula, and that the British government should not support the existence of faith schools at all. "Culture and beliefs can be transmitted at home," they write on their website.[29]

Yet despite the innumerable acts and customs of oppression people force on others in the name of their religious beliefs, it is unclear whether any atheistic movement would be a superior replacement. Members of atheistic movements frequently refer to Judaism and Islam as examples of explicitly sexist religions. Regarding Islam, many people oppose the burqa—the head covering Islamic women wear; regarding Judaism, people oppose the wigs women wear to cover their hair. Many people attack religion generally because it is sexist in doctrine; others note that some people justify rape and forced marriage through religious doctrine.

New Atheism, which started in the mid-2000s, was led by well-known academic and cultural figures including Richard Dawkins, Christopher Hitchens, and Sam Harris. A lot of established academics signed up to the movement, which espoused rationality and the critical investigation of superstition and religion. These people and their followers were explicitly anti-religious, and took advantage of the growing popularity of the Internet to make their beliefs widely known. This new set of beliefs entailed an explicit rejection of the existence of divinity colored by aggression toward the irrationality New Atheists (and humanists) believe is intrinsic to religiosity.

You would think this free critical thinking and focus on rationality would align well with the feminist movement, which espouses women's rights and equality with men. Since the persecution of women in major world religions remains a popular weapon in the atheists' arsenal, feminism should be fundamental to movements like New Atheism.

Instead, however, less than a decade after atheism's resurgence, stories started emerging that suggested otherwise. In 2014, the journalist Mark Oppenheimer published an exposé detailing misogyny and exploitation of women among New Atheism's respected members.[30] The article's major focus was Michael Shermer, an academic and founder of *Skeptic* magazine, who was, in 2008, an important figure among atheists—a free thinker who gained celebrity for his writings about rationality and the dangerous, exploitative practices of religions. Oppenheimer gives a detailed account of how Alison Smith, a twenty-six-year-old atheistic activist, was invited to a party by Shermer (then in his mid-fifties).

Shermer allegedly told her about the articles they could write together before plying her with alcohol (while, Smith noticed, hiding his own drinks under the table) and bringing her back to his hotel room for sex. "At a conference, Mr. Shermer coerced me into a position where I could not consent, and then had sex with me," Smith said, as quoted in a blog post by the biologist P. Z. Myers.[31] Shermer proceeded to lie to others about what happened that night; he even contacted a friend of Smith she'd called when trying to find her way back to her hotel.

Sexism is, it turns out, common among atheists. Women report being harassed at conferences. Richard Dawkins posted a video on social media mocking feminists.[32] Christopher Hitchens wrote an article in 2007 titled "Why Women Aren't Funny," arguing that men have an evolutionary need to be funnier than women. "Women have no corresponding need to appeal to men in this way," he wrote. "They already appeal to men, if you catch my drift."[33] Sam Harris, who has said in the past that people can be killed on the basis of what they believe alone, claimed that the "critical posture [of his atheism] . . . is to some degree intrinsically male and more attractive to guys than to women."[34]

Religion, in other words, is just one means of oppressing women. Atheism does just fine, too, and moreover the free thinking and rationality embedded in its doctrines give some people, like Michael Shermer, the moral credentials to behave no differently than any misogynistic religious fanatic.

Yet even though religious bodies are often the focus of discussions concerning oppression and exploitation, these problems are not restricted to them. In 2016, just as the Democratic party experienced a brutal defeat in the U.S. presidential election, Audrey Gelman, a well-known New York socialite and former aide to Hillary Clinton, launched a new company called The Wing. The underlying idea was to build a "safe and inclusive space for the advancement of all women and non-binary individuals."[35]

Gelman founded the company on strong foundations—intersectional feminism, equality, and racial justice—intending to empower like-minded women to work together and cooperate. Yet after the Covid-19 pandemic hit in 2020, The Wing didn't, it turned out, look after the lower-level employees working in menial roles. Women of color who had worked at The Wing's various locations came forward saying they'd been laid off, badly treated, and underpaid. Gelman resigned; the company was sold.[36]

Many organizations founded with altruistic intentions eventually develop exploitative practices, and anyone can be a victim. The actor Danny Thomas, who struggled financially in the 1930s in the United States, prayed to the patron saint of the desperate, St. Jude Thaddeus, for help. After reaching success, Thomas founded the now-renowned St. Jude Children's Research Hospital in 1962, which promises to treat children with cancer for free.

Desperate people from across America seek help from St. Jude Hospital, which is consistently ranked in the top ten pediatric cancer centers in the country. Yet despite the hospital's extensive public relations and fundraising work, not all parents find all of the financial relief they need. St. Jude clinicians treat children for free—but many other expenses that their terrified parents face, such as travel costs, hotels, or help with mounting personal bills following breaks from work, are not covered.

In 2021, *ProPublica*, which runs in-depth journalistic investigations into large institutions, reported that more than one hundred families of children treated at St. Jude had started GoFundMe pages to help with costs related to care.[37] And those who drove less than five hundred miles to get to the center were not given funds for a hotel along the way—including one mother of a child with cancer who St. Jude Hospital determined would need to drive only 491 miles (the mother's calculation, confirmed by *ProPublica*, was 530 miles). The hospital has, however, since said it will cover hotels for families driving more than 400 miles to reach it.

The reasons for these rules are not to do with any struggle for capital. St. Jude raises more than one billion dollars per year—two billion in 2021—receiving support from celebrities like the actress Jennifer Aniston. The surpluses run at over $400 million per year; the hospital's cash reserves exceed $5 billion. And yet despite the struggles despairing parents face, its executives, including the CEO, James R. Downing, have annual salaries exceeding $1.5 million, according to Charity Watch—a high reward for those doing the work of the patron saint of desperation.[38]

These cases are approximations of a broader possibility: institutions—and people—may maintain a good reputation even if they stop living up to it. Much as some organisms hide by blending in with their surroundings in a process called crypsis, so too can people hide behind reputations: we can call this social crypsis. The Wing and St. Jude Hospital are borderline cases—the latter does a great deal of good for people in need. And while most employees at any given institution may not intend to exploit others, it takes only a few powerful people to subvert the organization's ethos. What we should be afraid of are institutions that effectively exploit people without anyone's knowledge. That is invisible rivalry in the modern world.

This is not to imply that all people running reputable organizations intend to defraud the public. But a good reputation can provide the opportunity for exploitation. The problem is broader than individual organizations: whole ways of thinking provide these opportunities, too.

Take the idea (or myth) that more sophisticated technology improves people's lives. Cryptocurrency is one case where many people make this claim. Even some anthropologists agree: if we can track the exchange of money through a technological platform that cannot be infiltrated, we can prevent fraud and solve a huge amount of crime in the financial world.

This is unlikely. Instead, and I think predictably, cryptocurrencies offer just another opportunity for exploitation. The journalist Zeke Faux writes that Tether, one cryptocurrency that operates using a similar technological foundation to Bitcoin, is being used to run a human slave-trafficking ring in Cambodia. People, Faux says, arrive to a large complex after being promised work in the area. Instead, they are told they cannot leave, their phones and passports are taken, and they are forced—on pain of torture or death—to contact random numbers across the world to bait people into investment scams. Yet because of the invisibility people gain by using Tether—identification isn't necessary for using it—police can't track criminals using the currency. "It was hard to see how this slave complex could exist without cryptocurrency," Faux wrote about his investigation. "Crypto bros like to claim they were somehow helping the poor. But it seemed none of them had bothered to look into the darker consequences of a technology that allowed for anonymous, untraceable payments."[39]

If a Claw Is Caught . . .

Each of these cases, and there are innumerable others, creates the illusion that institutions deserve blame for the exploitation of people. This is not the case: people deserve the blame. Institutions, instead, are one of the means of exploitation in human societies.

It doesn't matter whether we're talking about religions, companies, charities, industries, academic centers, or any other organization made up of people. There is not and cannot be anything intrinsic to an institution that makes it exploitative except the people it comprises. Humans are, among other things, fundamentally creative: because of our long history of cognitive evolution, we've developed the ability to imagine solutions to problems we face.[40] This allowed us to develop new tools and behaviors for surviving in new environments across the world. But it also allowed us to navigate our social circumstances politically—to hide our intentions as much as to share them. It allowed us to be invisible rivals.

Coupled with an evolved predilection for pursuing capital, our intelligence, derived perhaps ironically from hundreds of thousands of years of social living, is the perfect tool for exploiting those around us. Psychopathy—that maligned disorder we both fear and enjoy reading about—is just the extreme version of that cold coupling. Psychopaths can navigate our social infrastructures better than others, achieving their ends at any moral cost.

Yet while people with psychopathic qualities disproportionately occupy positions of power in society, particularly corporate society, we should acknowledge that the traits they represent—those Machiavellian abilities that permit their use of others for selfish ends—are common to everyone, at least to some degree. Capital maximization is the social equivalent of a gravitational force: it pulls at people because it is essential for evolutionary success. We cannot reproduce without some combination of resource, social, and embodied capital, and so we rely on our intelligence to get it.

The markings of that force are everywhere around us, and at their core explain any example of exploitation by and within human institutions. In a future of healthcare report by the financial services company Goldman Sachs in 2018, the authors asked: "Is curing patients a sustainable business model?"[41] Yet in everyday conversation, anyone asking a question like this would be labeled a psychopath. It implies placing financial interests in treating chronic illnesses with expensive medications, like some forms of leukemia, over an immediate cure. It implies what many of us fear: that large corporations are more interested in addicting people to long-term treatments than to preventing illness.

The question is a marker suggesting that even if the analysts who wrote the report are not themselves psychopathic, they represent a psychopathic way of thinking embedded in modern life. The same market effects that permit a person to accept ten euros for a mouse's life permit corporate analysts to write about the future of healthcare as if people are cattle that should, for the sake of capital growth, be hooked up to intravenous drugs for the latter parts of their lives.

There is also no reason to assume anyone writing a report like this, or anyone part of an institution that exploits others, is a bad person. Our embeddedness in a global market economy, our ability to disguise from ourselves, through rationalization or self-deception, the costs to others of our actions, and our spread of guilt across corporations, industries, and political structures give us the ability to say and do things that, when viewed from the perspective of an investigative journalist, look psychopathic. That means only that we tolerate bad behaviors—and that some people, like cancers, behave more selfishly than others. And like cancer, it only takes one cell to disrupt a whole organism's functioning.

Many of my colleagues nonetheless contend that successful societies consist of prosocial people, and consequently that cooperation is our default in interactions with others. And although we may be happy to help others, we're unlikely to do so if the cost of helping prevents our achieving a goal we consider important. You might be happy to give a homeless person some money, but not to invite that person to live in your spare bedroom (or spare house). You might talk about equality in education, but you might also pay for a tutor or an expensive education if you can afford it. Getting ahead comes first; giving back is the afterthought.

That relentless pull toward self-interest is the power Tolstoy alluded to in his disturbing play. The social forces around us give us the means to ignore our better judgments to achieve our goals. And those who are the best at this will hide their intentions from others, and maybe even from themselves.

This is the view of human nature we need to adopt: it will inform how we combat the multitude of threats we face. To the degree that we believe we are inherently cooperative, we will be complacent and allow the continuing exploitation, through institutions and otherwise, of the people and world around us. Yet combating the power of darkness requires, first, acknowledgment—and the will to break the often unethical norms that pervade the modern world.

6

Electrification

Light's all very well, brothers, but it's hard to live with.

—Mikhail Zoshchenko, "Electrification"

In September 1945, from Nuremberg prison, Albert Goering wrote the following to U.S. intelligence: "Herr Oberst, I am trusting in my belief in God and your humanity and the American sense of justice, which probably wants to avoid injustice."

If you recognize Goering's last but not his first name, yes, Albert was related to Hermann. For those who do not: Albert Goering was brother to Hermann Goering, head of the German Luftwaffe during World War II and founder of the Gestapo. He was also for much of the war considered Adolf Hitler's right-hand man—and would himself have been named führer in the event of Hitler's death.

Yet Albert, unlike Hermann, was not a National Socialist. He had never, according to documents from U.S. intelligence, been a member of the Nazi party. Instead, he used his brother's influence to defy fascism: he refused to greet others with the required "Heil Hitler"; he protected elderly Jewish women from German soldiers. Most important, perhaps, he helped dozens of Jewish people to escape death—much like the more famous Oskar Schindler, but without the adulation. U.S. officials didn't believe his story, and it was only when evidence started gathering and people testified about Albert's morals, hatred of the Nazi party, and actions to preserve human life that he was found not guilty of being a war criminal.

Whereas Oskar Schindler received support from the Jews he saved during the Holocaust and was the subject of Steven Spielberg's famous movie *Schindler's List* in 1993, Albert Goering spent his remaining years working at a construction firm in Munich. He died in 1966, in obscurity and estranged from his only daughter, who lived in Peru, and who wrote him unanswered letters until her tenth birthday.[1]

Many people have speculated about Goering's intentions in defying his brother and the Nazi party. He didn't look like his brother, leading some to speculate that he was the son of his half-Jewish godfather. It's also possible he enjoyed the thrill of tempting Nazis: who, after all, would challenge the brother of the founder of the Gestapo?

We may never know. But the tale of the Goering brothers shows that there is much more to a person's ethical life than conformity or a desire for prestige. Being altruistic is difficult—it requires sacrifice, though its possibility does not mean that altruism is any person's defining feature. It is proof, instead, that we can resist even the worst cultural contexts without any promise of benefit. The secret of our success, to steal a phrase of an anthropologist's, lies there: our will to defy our genes and our cultures, no matter the price.

Albert Goering's 1939 passport photo (from Adam LeBor)

The Costs of Conformity

Ron Jones, a former teacher living in California in the 1960s, was, like many others, unable to understand how people living in Nazi Germany could have tolerated the atrocities the regime committed. His answer was to start a movement, the Third Wave, in the high school class he taught.

The movement lasted five days. Over this time, he encouraged members of his class to adopt a slogan: *Strength through discipline; strength through community; strength through action; strength through pride*. The Third Wave quickly expanded out of his classroom, and by day three more than two hundred students had joined. Tribalistic behaviors started emerging, and students began to report on each other when they breached Jones's punitive rules.

By the fifth day, Jones had led Third Wave followers to believe that they were involved in a national movement for which they would support a U.S. presidential candidate. He brought them to the school's auditorium—and at this climactic moment, he told them the Third Wave was an experiment he had concocted, and its success was a harsh lesson to them all about how fascism takes hold.

Jones, who still lives in California and teaches worldwide about his and his students' experience, was kind enough to speak with me online about the Wave. "I didn't have a plan," he said. "I just saw what was happening each day and improvised about what to do next."

Jones told me that he knew implicitly which students would ask difficult questions and create resistance to his movement—and perhaps unsurprisingly, these were the students who sat in the front row. "You always notice the students in front, because they think like you and ask good questions," Jones said. "You also notice the students in back who come in late, because they are rebellious and interesting. But you don't notice the ones in the middle. And when you remove the students in front, the middle moves up—and they became the most engaged."

The Wave created a feeling of shared identity among its members—high school students growing up during the Vietnam War, who felt the tension of change in their surroundings. This mirrors recent work in anthropology that shows when people, including children, move together synchronously, such as in a dance, they are more likely to be trusting and effective at a cooperative task than are others.[2] But coupled with the "anti-Wave" signs that started popping up around the high school, those with newfound in-group solidarity at once had a leader who understood them and an enemy to fight. "Autocrats undoubtedly create that feeling intentionally," Jones told me.

The Wave has inspired a huge amount of media attention, writing, art, and film. As with psychopathy, people share an interest in understanding what they fear is within them: the ability to override their sense of ethics and justice, and conform to oppression of a group that can't defend itself. Jones's experience suggests, however, that anyone can be perpetrator or victim, and the difference is merely a matter of opportunity—and belief about what is right.

Researchers working on how cultures evolve share this interest—albeit perhaps for more academic reasons. This growing body of work is helping to explain how and why people adopt the beliefs they do, and how those beliefs fit in with a larger cultural framework, which is the lens through which we understand the world.

Two patterns have emerged in this area—known now as cultural evolutionary studies—over the past several decades. First, these researchers suggest humans share a bias toward conformity. We are inclined to adopt the beliefs and norms of the majority around us. If you are born into an Islamic society, for instance, you are likely to identify as Islamic. Second, we have a bias toward prestige: we imitate and tend to share views with people we believe are successful, interesting, attractive, and so forth. For people working on how cultures evolve, these are thought to be foundational qualities that we evolved to share.[3]

Of course we adopt views in other ways, too. Yet because growing up necessarily involves adopting a worldview encompassed in a language, we need to rely on heuristics for how we absorb the huge amount of information our cultures throw at us. Our childhoods are so long because human culture is so complex, and consequently requires so much time to absorb. We are born blank slates, culturally speaking, and spend the first eighteen or so years of our lives becoming an expert in local social norms.

This near-total reliance on learning from others, or what's called "cultural transmission," leads to some strange findings in studies of comparative psychology. In one experiment, researchers compared how well human children and chimpanzees performed in a task that involved obtaining a reward from a box. Opening a box was a simple affair, and researchers showed both the children and the chimps how to do so. For both species, though, the researchers also performed a ritual before opening the box: they moved their arms in a complex way that had nothing to do with obtaining the reward.[4]

When it came to opening the box and getting whatever was inside, the chimps performed well—they undid the box's mechanism and took their reward. Human children, interestingly, took longer: they imitated the ritual movements, even though they had nothing to do with opening the box. The cultural elements of the exercise were of equal importance to the humans as were the box's mechanisms; the chimps were interested only in the fastest way to obtain the reward.

This difference is harmless in a reward-based experiment, and researchers repeatedly point to our marked predilection for cultural transmission when attempting to explain why humans have been so successful in our social conquest of the earth. The biases in how we take in information have drawbacks, though. The psychologist Nicholas Humphrey pointed out twenty years ago that biases for conformity and prestige have the unfortunate drawback of rendering us vulnerable to autocracy. If personable, attractive, and deceptive people like psychopaths and narcissists use their social abilities to become successful, and we in turn become drawn by their prestige, we have the ingredients for being tempted to follow a dangerous leader. "For the fact is—the worry is, the embarrassment is—that human nature may sometimes be not just a collaborator but an active collaborateur with the invader," Humphrey wrote in 2003.[5]

This process explains a lot—why so many people are drawn to cult leaders, for instance, and why psychopaths are invariably overrepresented in positions of power in our cultures. In more extreme cases, like Nazi Germany, Adolf Hitler represented not only the subversion of government for dictatorship, but also the will of a society enslaved, through biases we evolved, to his insanity. "Not theorems and rules ought to be the rules of your Being," the philosopher Martin Heidegger said in 1933.[6] "The Führer himself personifies German reality and its law for today and the future." Through mastering the biases of conformity and prestige of the German people at a critical moment in history, Hitler created the impression that he embodied their will. To anyone familiar with political events in the United States from 2016 onward, the rise of Donald Trump should be worrying. His refusal to admit his defeat in the presidential election of 2020, to name just one example, was worrying because so many of his followers—among them other important political figures—were so ready to believe him, and to insist that he had not lost.

Humphrey is not alone in his concern that autocracy is a constant threat to open societies that champion individual freedoms. The late economist Gordon Tullock developed several models over the course of his career suggesting that open societies inevitably move toward autocracy, unless sufficient guards are in place to prevent this regression. He noted in the early

2000s that autocracy—and specifically dictatorship, which lacks a process of succession—was the most common form of government worldwide, with an increasing number of democracies devolving into autocracy in regions such as South America.[7]

Unlike Humphrey, who focuses on evolutionary processes, Tullock was writing from the perspective of political science. He developed mathematical models to show that the ways society invests in its own security can lead to more centralized power, and eventually, autocracy, in the effort to preserve efficiency when faced with resistance from within our cultural groups. In the language of biology, we would call this a "basin of attraction"—a metaphorical black hole toward which evolutionary change is directed. Democracies, to combine Humphrey's and Tullock's views, are drawn toward dictatorship because of our evolved biases and, ironically, our efforts to protect them.

This pull is not, however, inevitable. People can be involved, cognitively speaking, with what social norms they adopt. We have emotional reactions to atrocities that go past what we're brought up to think or feel. And it's that characteristic of being human that gives us the power to change what is normal: we can design society to represent the views we think are right, not just to confirm what we are told.

There are, of course, effective and ineffective ways of doing this, as well as right and wrong ones. Since the rise of artificial intelligence, for example, people have argued that these tools are likely to damage the well-being of people across society, from replacing jobs to threatening human existence. "Robots are coming for your jobs" is a mantra of people invested in the world of technology. And with the advent of large language models—tools like ChatGPT that replicate, sometimes convincingly, human interaction—people like writers who thrive on representing human sociality feel that their livelihoods are in danger.

This issue partially motivated the strike of the Writers

Guild of America (WGA), which represents more than 11,000 screenwriters, in 2023. The writers were worried that ChatGPT, or its relatives not yet released publicly, would be used to replace them—a fear echoed by media coverage suggesting large language models are as creatively capable as any human.

After much hesitancy from the Alliance of Motion Picture and Television Producers, the two groups came to an agreement. That agreement was striking, according to a journalist from the *Financial Times*, because it suggested that—against claims many researchers and politicians are making—it's possible to regulate artificial intelligence.[8] First, members of the Writers Guild can now stipulate whether they want AI tools used in projects they are working on, and second, production studios must inform writers whether any materials they share with them were developed using AI. This gives writers the power both to veto AI's use, and to ensure that their intellectual property—the words they write down—remains their own.

From an anthropological point of view, this reflects more than a victory for an underpaid group concerned about the future of their professions. It shows that even in a country such as the United States where there is widespread inequality of resource and social capital, a unified group can aid in the regulation of evolving technology. These bottom-up approaches are not usually used to develop policies for large industries, and before the WGA strike agreement, tech companies were requesting that politicians regulate these new tools in a top-down manner.

The *Financial Times* journalist Rana Foroohar notes that the International Brotherhood of Electrical Workers was similarly pivotal to the regulation of electricity when it was introduced in much of the southern United States by the Tennessee Valley Authority in the 1930s. Consultation with the workers, who knew their craft at the ground level, was essential for the success of this project. Contrast this with Vladimir Lenin's goal for electrification across the USSR. "Communism is Soviet power plus the electrification of the whole country," he said in 1920.[9]

His top-down approach and refusal to acknowledge the myriad other social problems in post-revolutionary Russia led the Soviet writer Mikhail Zoshchenko to compose his short story titled "Electrification."[10] Lenin's plan, in Zoshchenko's view, merely allowed people to see the awful conditions they were living in: poverty, grime, and a lack of power to change their circumstances. "Light's all very well, brothers, but it's hard to live with," he wrote.[11]

More generally, the late Nobel laureate and economist Elinor Ostrom argued that top-down approaches by governments were likely to fail in achieving their aims. Consulting the people policies are likely to affect is, in her view, critical for success and the well-being of those they affect. This is essential when facing a widespread problem known as the tragedy of the commons. These tragedies, first identified as such by the ecologist Garrett Hardin in 1968, occur when people collectively overuse a resource that's in limited supply.[12]

A common example of this is supply in fisheries. Each year, people fish for food worldwide, and historically fishing stocks have reproduced sufficiently to keep enough fish in supply. Yet as populations grow and more fish are required to feed an ever increasing number of people, overfishing—which prevents stock replenishment through breeding—becomes an increasingly serious problem. And worse, to the degree that a single fisherman scales back his fishing, others will surpass him economically. So, Hardin concludes, all fisher-people have a similar, rational interest in maintaining their personal hauls, with the long-term consequence that, after many years, there are no fish left to catch.

These tragedies of common behavior constitute what many

people in psychology and economics call a social dilemma. We each have a personal interest in maintaining—or increasing—our relative resource capital, but where there is limited supply, our interests collectively lead to a net capital fall for all. Otherwise put: we are each better off if everyone reduces capital gains in the short term, but, rationally speaking, none of us has an interest in doing so.

The consequences of these tragedies can be severe. All of the resources humans rely on are becoming scarcer, and that increasing scarcity is not limited to intuitively obvious examples like food and drinking water. We also rely on natural resources like arable land and trees, and artificial ones like irrigation systems and, perhaps surprisingly, the Internet. And as the world continues to warm because of human carbon emissions, many areas will become uninhabitable within a few generations, depleting the amount of livable space needed for a rapidly growing overall population.

Ostrom identifies two features these resources—called common pool resources—have in common.[13] First, people, groups, or institutions can exclude some people from accessing them, though it may be costly to do so. Farmers buy land that delineates where only they can grow crops or only their cattle can graze. Villages can dictate who is allowed access to local drinking water. And countries—including modern-day Russia and China—can prevent the public from obtaining unsanctioned information by blocking access to the Internet. One report in 2022 suggested, to this point, that governments in many low- and middle-income countries are increasingly turning to Internet blackouts to prevent the public from accessing information about demonstrations, military coups, elections, violence, and even religious holidays.[14] Where knowledge is a threat, the exclusion of some groups from information channels can be an effective tool for suppression.

Second, common pool resources must be subtractable. Their use by anyone must, at least over time, reduce the amount available for others. Fishing reservoirs are a good example, but electricity, clean water, and the World Wide Web are also subtractable. For the last of these, its use does not leave less Internet for others—but rather the energy required to transmit the information stored online is not limitless. This does not, for now, affect wealthier countries, but where Internet access is less widely available, as again is the case in many low- and middle-income countries, energy blackouts can prevent people from obtaining critical information about health crises and political unrest.

Problems about how to manage common pool resources pervade human history. And because exclusion and rivalry (the state of being subtractable) are interrelated—it's harder to exclude someone from a limitless resource—the evolution of social norms has been a critical element to solving these ancient dilemmas. Yet ignoring or attempting to artificially replace these norms can be ineffectual at best or disastrous at worst.

Ostrom points to farming systems in Nepal as an example. Before government intervention, local farmers collectively managed the Kathar irrigation systems in Nepal's Chitwan District.[15] The farmers had longstanding rules and norms about the systems' management, which, presumably, had evolved over successive generations to maximize consistent water availability for crops. When the government-backed engineers installed technologically sophisticated replacements without consulting farmers about their norms for management, the consistency in water availability dropped. The farmers' older systems, which functioned using mud, stone, and trees rather than concrete and steel, were more effective—they had the backing of evolved social norms.

Analogous problems plague fishers in the Amazonia area

of Brazil. In 2005, the Brazilian government instituted a policy called the *defeso*, which constitutes a closed fishing season while compensating fishers for consequent lost profits. In theory, the defeso would support fishers while allowing stocks of fish to reproduce, solving the social dilemma from the top down. Yet in practice, the policy is not enforced, leading to continued fishing during the ostensibly closed season. Moreover, the authors of a recent research paper argue that the policy creates an incentive for an increased number of fishers overall through a subsidy given during the closed periods. The defeso is "truly a formula for disaster," the authors wrote, and creates a greater incentive for more fishing in the short term, resulting in reduced stock over time.[16]

Poorly enforcing rules and policies, Ostrom argues, is generally devastating for collective aims, and moreover occurs more often with top-down approaches. Insofar as people watch others breaking rules for economic games without being punished, those behaviors are likely to be spread—a consequence predictable both by appealing to how culture evolves, and in a more rudimentarily obvious way. Rationalization is easier to the degree that rule-breaking is normal.

Enforcement, then, may be one element of a larger institutional framework needed to solve the myriad social dilemmas people face at the local and global levels. Yet at the local level, institutions must be designed to accord with the norms of the populace they aim to support. This implies a need for a diversity of cultural institutions reflecting the diversity of cultures, which in turn reflect the diverse ecological universes we inhabit. In redesigning norms we must accommodate those layers of diversity, but we must also respect that the problems we face today are different from those of our past. Our social dilemmas are increasingly intertwined as our cultures become more connected, and thinking about how we can scale up from the local to the global is a necessity in redesigning what is normal.

The Free-Rider Problem Revisited

Understanding how and why social norms evolved is the first step in ensuring we do not eliminate, accidentally or otherwise, our ancient strategies for thwarting free riders in society. We've seen that when we evolved in smaller groups, we designed over many generations the social norms that governed the distribution of the resources we relied on. People in varying positions in society vied for power, and the norms that emerged, such as the Golden Rule, help to maintain a trend toward equality, balancing out our more selfish dispositions.

The free-rider problem, however, persisted, and continues to persist in modern societies. Some people free ride overtly and don't, for example, file their tax returns in large countries like the United States. Others hide their cheating tendencies, and take for themselves only when they believe they won't be caught. More than one hundred thousand people and business entities avoided paying taxes by hiding money abroad, as the Paradise Papers, leaked in 2017, revealed.[17]

These and others who hide their free riding are our invisible rivals. And because they—whoever they are—are effective at hiding their duplicity, it's difficult to track who is taking advantage of society's common goods without giving anything in return. Invisible rivalry leaves a signal only in its effects.

Yet in modern society, hiding resources to the extent seen in the Paradise Papers is possible only because we rely largely on intangible assets for monetary exchange. Risk-pooling systems work effectively in smaller, preindustrial societies because people cannot usually hide their assets. This is not true of a globalized economy: people can and do misrepresent their means and not only avoid paying taxes, but invalidate the effectiveness of social norms around risk pooling. If I do not know how much money you have, and you do not know how much money I have, we will be less likely to help one another: either of us could lie and say we need more than we do, or that we don't have enough to help the other.

But this twin problem does not suggest that need-based sharing systems, where people help others only when help is needed, can't work in the presence of free riders. Anthropological work showed, for instance, that exploiting the system by asking for resources when none are needed doesn't lead to a need-based sharing relationship falling apart. Maasai pastoralists living in Tanzania told anthropologists that they share a small number of cattle with people who didn't need any to survive. Yet the researchers also found that where there was a genuine need, generosity was greater.[18]

The authors created a computer model that confirmed their findings. Cheating didn't have a huge impact on the sharing system so long as people weren't stingy—so long as they did not, in other words, refuse requests for resources from others. Free riding in the second sense—requesting resources that weren't needed—did not lead the system to break down.

The findings have broader implications than just for small pastoralist communities. Many modern politicians argue that because some people free ride on public benefits, we should restrict those benefits to smaller groups. Giving to the unworthy can lead to wasting taxpayer money, as fraudulent business requests during the Covid-19 pandemic showed. The U.K. government, for example, provided about £1.65 billion to organizations that falsely filed for pandemic-related relief.[19]

Yet economists have repeatedly shown, in laboratory experiments and in analysis of real-world data, that free riding is, at least to a degree, unavoidable: some people will always ask for resource capital that they don't need. This doesn't mean, however, that we shouldn't help those who are in need—we just need to accept, as the Maasai seem to, that free riding will happen. Planning and budgeting for this in policymaking is one way to avoid the political rage that surrounds it. Any funds saved by catching free riders will then be a bonus.

Analogous reasoning applies to the international level. Just because some countries misrepresent the effectiveness of their sustainability policies doesn't mean that everyone should abandon their climate goals. Instead, again, free riding should be accounted for in models that aim to predict reductions of carbon emissions. We know free riders are among us at every level of society—and pretending otherwise can make our own goals unrealistic, and worse, appear hopeless.

Relatedly, recognizing free riders may be more important than punishing them. "Throughout history, rulers often beheaded criminals, including tax evaders, in the town square," Jonathan Weigel, an economist, recently said to me. "But modern states have learned that in fact they are better off detecting and exposing more avoiders, rather than heavily punishing fewer of them."

This suggestion mirrors how some people appear to play games in laboratory experiments where they can choose to behave honestly or dishonestly. One study from 2008 showed that where the risk of being caught was high, people tended to behave more honestly—even if the economic punishment for being dishonest was low. Worries over being perceived as a cheater, or punished for being a cheater, drive much behavior, as many empirical studies have shown.[20] These concerns may even drive tax-paying behaviors, such as Weigel's work in the Congo shows. "Although the probability of enforcement of the property tax among average citizens is vanishingly low, the tax authority manages to sustain surprisingly high enforcement beliefs by occasionally locking delinquent commercial properties in a public facing way," he told me. "'Locked by the DGI' reads the sign on properties in downtown Kananga whose front door is locked by a large exterior facing padlock."

The mirror tendencies to free ride where possible, and to avoid being perceived as a free rider, are ubiquitous across the layers of any society. That does not mean that all people will free ride in the absence of legal or social repercussions for doing so. But it does suggest—in line with experimental research—that a proportion of any group of people will free ride where possible. There is nothing about a person's identity, whether we talk of resource capital, status, class, education, or ethnicity, that predicts economic behavior at the individual level.

Many people harbor a bias against those who are homeless, for instance. Studies in North America have suggested that the public doesn't trust homeless people to manage their money wisely. Given that drug use is common among these people, a common expectation is that a homeless person will spend money on drugs.

Yet a study from 2023 suggests that this isn't the case.[21] Researchers in Vancouver gave fifty homeless people a one-time unconditional cash transfer of 7,500 Canadian dollars. There was no stipulation about how the participants should spend the money.

They tracked how the money was spent over one year, and compared spending trends with sixty-five people who weren't homeless—who were also given a $7,500 transfer. There was no difference after one year in spending on drugs, alcohol, or cigarettes between the two groups.

Public sentiment about homelessness is therefore highly likely to be dubious in foundation, and instead reflects a bias we should collectively try to overcome. The reasons for this are not only founded in faulty logic, let alone ethics, but economics.

Per person, homelessness costs the United States more than $5,000 annually in health and social care—and where a person has mental health issues, which is common in this population, the costs can exceed $80,000. Yet the Vancouver study's findings suggest that the unconditional cash transfer—even of a sum that is only 12 percent of the city's average income—can create a net savings for local governments.

Although it's convenient to rationalize about inequality using the illusion that cheating and norm violations increase in frequency from the top to the bottom of social orders, the truth is less convenient for those in power. Violations are everywhere, because people are everywhere. And at the societal level, refusing help to those in need based on assumptions otherwise is unjustified.

The Contracting Circle

The philosopher Peter Singer wrote in 1981 that, as society advances, we should broaden the circle of people we help.[22] Among preindustrial groups, such as the Maasai, a person's circle might be limited to family members and those they have sharing relationships with. By contrast, in a large, globalized society, people should—I think in an ethical sense, on Singer's view—value the needs of more people, and animals, they aren't connected to. We have greater means, so we have a duty to help more people who are in need, regardless of whether we have a familial or social connection to them.

Strangely, and despite whether Singer is correct, the opposite economic trend seems to be true. The *Financial Times* reported in 2023 that Claridge's, a luxury hotel in London, had opened a newly refurbished penthouse.[23] The rooms included revolving sofas, an outdoor lap pool with a glass pavilion containing a Steinway grand piano, and seventy-five works of art by Damien Hirst. The room costs, according to the article, £60,000 per night, equivalent to about $80,000 in U.S. dollars. (For reference, the median annual income in the United Kingdom was about £32,000 in 2022.)

This is an extreme example, but inequality, at all levels, is greater among large, industrialized societies than any others. And the problem only seems to be getting worse. Data from the World Economic Forum show that over the past decade, the richest 1 percent of people gained more than half of all wealth created worldwide.[24] Between 2020 and 2021—that is, in the midst of the Covid-19 pandemic—the richest 1 percent gained 63 percent of all new wealth. The poorest 90 percent of the world, by contrast, gained just 10 percent of wealth created in the same year. "The 'average' billionaire has gained roughly $1.7 million for every $1 of new wealth earned by a person in the bottom 90 percent," the authors of the report wrote.

Yet inequality is not limited to money, and certainly not to money within the present generation of people. Globally, each year, more than three million babies die from insufficient nutrition, accounting for nearly half of all pediatric deaths. It's estimated that more than 200 million children are undernourished, which leads to stunting—low height for age—and in some cases, muscle wasting. And when you add being overweight to the equation, nearly half of the roughly eight billion people alive today are malnourished, whether that's from too little food or too much.[25]

Malnutrition of both kinds is linked with chronic health issues, and the poorest in society predominantly suffer from them. People who are overweight or obese are at a higher risk of cardiovascular diseases and cancer—the two leading causes of death in developed countries. And people who are undernourished not only suffer in their own lifetime, but can transmit their suffering to their children. Research shows that fetuses that do not receive sufficient nutrition before they are born divert energy from organs like the pancreas to the brain, which requires a huge amount of nutrients to grow properly.[26]

People whose bodies grow in this way—whose life history, to use the language of biology, is altered by poor nutrition in utero—suffer a long-term price for this trade. They cannot tolerate high-energy diets, for one thing, and when they take in more food than their bodies can process, they are at higher risk of type II diabetes, which is in turn linked with risks for cardiovascular disease, blindness, open wounds resulting in lost limbs, and cancer.

Differences in resource capital can translate more directly into differences in life expectancy, too. One study showed that if you travel by train in London from wealthy Westminster to poorer Canning Town much farther east, you will see populations with consecutively lower life expectancies. For each station you stop through, the average man will have a 0.75-year worse life expectancy, while for each woman the drop is about 0.5 years. The distance is less than eight miles, but the overall drop in expected years of life from birth is more than six years among men, and nearly four years among women.[27]

Education, which is linked with resource capital, corresponds with similarly stark predictions about a person's life. Data from the Brookings Institution in the United States show that by age twenty-five, people with a bachelor's degree can expect to live, on average, eight years longer than their less educated counterparts.[28]

Money, nutrition, and education are together the triad that predicts a person's life expectancy in the developed world. These qualities, representing different types of capital found in humans, reflect the struggle for power across cultures, not just the developed world. "The relationship between nutrition and health is deeply embedded in power relations," writes the anthropologist Jonathan Wells, who focuses on the relationship between food, health, and power across cultures.[29] Wells writes that capitalist systems create the unique circumstances in which we find both stunting and obesity among the poorest, least educated among us.

These struggles lead to corrosion in society over time. Economists have shown that greater inequality in mortality and resources is linked with lower public trust. Richard Wilkinson, a social epidemiologist, has pioneered this area of research, and has shown repeatedly that many Western societies have created what he calls a "culture of inequality."[30] These cultures normalize unequal relations, as in the United Kingdom where, in 2021–2022, more than two million people lived in a home that relied in part on food banks—while a single hotel room might be booked for nearly double the country's average annual salary per night.[31]

Wilkinson notes that aggression and violence are greater where inequality is greater, and where community support is lower. Male behavior, he says, is in particular more aggressive in these societies, and relatedly people start to blame vulnerable people in society for their problems. Religious minorities such as Jewish and Islamic people are some of Western societies' favorite scapegoats, but women and ethnic minorities are also targets.

The dynamics that lead to cultures of inequality are social, but they are mirrored in evolutionary biology. Research suggests that when individual animals adopt long-term strategies of stealing food—termed "scrounging" in the literature—from others, they grow larger and gain higher social status.[32] Over time, fixed strategies like scrounging can introduce dominance hierarchies into populations, which can in turn affect reproductive patterns. Larger males in some species, such as gorillas, can effectively defend their harems of mates, and are consequently more reproductively successful than others.

Scrounging takes many forms in the animal world. Some aggressive birds like gulls force more docile creatures, like puffins, to give over food they've caught for their young with threats of attack.[33] Brood parasites like cuckoos are also scroungers: they trick or force host birds to feed their young. What each has in common, however, is the guile or strength to take resources that others worked to procure.

Cultures of inequality operate in an analogous way, with different modes of exploitation in different societies, corresponding in turn to different cultural histories. Over time, and throughout the history of the evolution of human culture, people in dominant positions have adopted ways of sustaining power. This may not involve aggressive bullying, which we see in bird and some mammal populations, but rather the effective use of intelligence and language—and eventually, military power— to solidify hierarchies. In the sixteenth century in Europe, for example, many monarchs claimed the divine right of kings— the notion that a deity proclaimed them to reign over their people.

In the modern setting, those with resource and power are similarly effective at controlling how others view them, according to my colleague Devi Sridhar. "You have an oligarchy controlling the narrative through ownership of major national and international publications—and this shapes what the public thinks," she told me. "Politicians are then under the behest of these papers."

Yet whatever its structure, any culture of inequality is founded in differential rankings that, at their core, stem from scrounging patterns. Over time, people become defined by their heritage, which in part constitutes their social status. "'Tis very certain that each man carries in his eye the exact indication of his rank in the immense scale of men, and we are always learning to read it," the American writer Ralph Waldo Emerson said.[34]

Otherwise put: control of the mode of exploitation leads to consolidated social capital and status, and consequently cultures of inequality. This may help to explain patterns of violence in cultures with social stratification: humiliation and disrespect are the primary reasons for aggression, according to research from the Center for the Study of Violence at Harvard University.[35]

The tentacles of resource inequality are therefore as diverse in form as they are damaging. And yet the trend is not, despite philosophical arguments about ethical duty, improving. This is probably another consequence of the social myopia most people inherited from their socially stratified worlds.

The frequent—but not necessary—shift from left- to right-wing thinking as people age in many Western countries illustrates this. As people graduate from university or their first jobs into higher-paying roles, their social groups change with them. Buying a house and sending their children to good schools become important motivations, requiring increasing resource investment in their own and their family's futures. So while it's easy when you are a student or a young professional with few expenses to talk about our expanding circle of duty to others, our circles often contract as we age. We talk about helping others less because we need to focus on the circles around us— and ensure that we keep up with the people we are surrounded by. I think this is what people mean when they talk about reality setting in. Yet that reality is—in reality—social myopia characterized by competition with the people around us. No one wants to fall down the social ladder, and increasingly costly resource investments are necessary to maintain the same social capital. The result of the contracting circle, however, is ever greater social inequality.

There is evidence, though, that this pattern is changing. Polling data from 2023 suggest that, unlike the generations before them, people born after 1985 or so are not becoming more economically conservative.[36] Housing crises in the United Kingdom and the United States are instead purportedly driving younger people to push for higher taxes on wealth. The limits of resource inequality may, for the millennial and Gen Z generations, some of whom aren't so young anymore, have been reached.

A common question, then, is how to address growing inequality both within and between countries. Some polls suggest that many people hold incompatible views about taxation and fairness in political systems. They want, analysts say, reductions in poverty and fewer people to be homeless. But they also want lower taxes. We can't have both, though: the social safety net costs money, and we need to fund it through higher taxes. Alternatively, we can lower taxes and reduce the efficacy of the state's ability to reduce the number of people living in poverty. Wanting both is, it seems, irrational.

Together with two co-authors, my colleague Daniel Nettle ran a survey that aimed to address this apparent irrationality. The study relied on a complex type of survey known as "conjoint," which explores how people rate policies relative to each other. It evaluates not only the binary questions of whether people prefer lower or higher taxes, or more or less poverty, but also participants' preferences rated on scales: some people might, for example, be willing to have slightly higher taxes for a great reduction in poverty.[37]

Nettle and his collaborators discovered that not only are people consistent about their views, they rated—on average—

reductions in poverty as the single most important policy to them. The surveyed group, which included 800 people living in the United Kingdom, even expressed a willingness to accept a 10 percent rise in taxes for a major national reduction in poverty. They also preferred wealth, carbon, and corporate taxes as means of funding these projects, rather than relying on government borrowing.

Research from anthropology supports increasing taxation as an intervention for reducing inequality. A survey of twenty-one small societies across the world showed that the amount of resources people can inherit from their families directly predicts inequality more broadly.[38] The transfer of resource capital, rather than the production of it, was the biggest statistical predictor of unfair wealth distributions.

To expand our circles, then, we need to address inequality by breaking the transfer of resources from one generation to the next. This does not require a social revolution, but rather representation of this perspective—and the perspective, it seems, of respondents to well-designed surveys—in politics. Instead of relying on billionaires to give their money away when they die, as signatories of the Giving Pledge have promised to do, we should collectively prevent the transfer of these enormous sums altogether through effective taxation.

Yet while many countries with high concentrations of wealth, such as the United Kingdom and the United States, have inheritance tax policies, there are frequently loopholes that permit people to avoid payment. The Organisation for Economic Co-operation and Development published in 2021 data illustrating that only 0.2 percent of estates in the United States, and 3.9 percent of estates in the United Kingdom, are subject to inheritance tax—compared with 48 percent in the capital region of Brussels.[39] It is not surprising, then, that among Group of 7 countries, the United Kingdom and the United States have the highest indicators of income inequality. Furthermore, the same indicator, the Gini coefficient, is much lower in Brussels.[40]

What's more, there is no indication that the hyper-wealthy have any intention of paying tax. In the Global Tax Evasion Report for 2024, the European Union's Tax Observatory noted that billionaires pay personal tax rates of about 0.5 percent in the United States—and close to nothing in France, which otherwise has a high income tax rate.[41] Shell companies allow these people to hide their wealth, leading working-class people to sometimes pay higher rates than they do. It is therefore suspect to rely on declarations of generosity in death from those who misrepresent their wealth in life.

We have looked to hunter-gatherer and foraging societies to see what drives inequality. And, consequently, we have the knowledge to reduce it. We just need the political will to enact policies that upend the modes of exploitation we have normalized and the cultures of inequality we allow to thrive. Otherwise, the dynamics of exploitation will continue much as they have done for the entirety of modern history: "Scrounging will occur whenever it is profitable and producers are unable to prevent it," Jonathan Wells writes.[42] If a system can be exploited, it will be.

The Trust Game

Growing inequality in contemporary times is likely to only worsen public trust. This will permit, by consequence, widespread disinformation campaigns to succeed, which may affect climate goals, public health programs, and political movements both within and between countries. Too often, as some of my colleagues who work on perceived catastrophic threats like artificial intelligence suppose, we ignore the underlying reasons for why we seem further away than ever from solving global issues. The underlying problem may be the fundamental lack of trust in society.

The free-rider problem, however, makes it difficult to find a solution. We can't always—or sometimes, it seems, ever—tell who deserves our trust. Trustworthiness is an unobservable quality, analogous to parasite resistance in some birds or fertility in most species. In both cases, animals evolve ways to signal these underlying qualities, as with color badges on birds that help to show parasite resistance. Trustworthiness in humans may work in a similar way. We can't just say "trust me"—we have to show that we are worthy of trust.

Yet unlike with the animal world, the costly signals we use aren't generally reliable. Most birds with low parasite resistance can't grow a large, energy-intensive patch indicating their qualities. A human can, however, fake indicators of trustworthiness. People can deceive others, and maybe even themselves, about whether they intend to help others, or why they give money to a charity. Some people follow social rules and pay their bills because they believe it's right to do so; others act in these ways because they don't have the opportunity not to. And yet the appearances will be the same.

An appeal to evolutionary theory can help to solve this puzzle. Birds can't fake badges of resistance because they are energetically costly. Cost in this case is a straightforward concept: birds without resistance don't have the excess energy to grow the badges. Yet with humans, trustworthiness is similar to the animal world—but what's difficult is establishing what costs in human societies entail.

My guess is the definition of costs is contextual to the circumstance—but it will always involve a sacrifice of capital, or at least the potential for capital. Much as the meaning of words arises from the contexts in which they are used, so do the costs of signaling social qualities vary from situation to situation. Many researchers, writers, and politicians talk about how resource and health inequality are critical problems societies need to address, for instance. Yet few people are willing—and I count myself among them—to reduce inequality themselves. No one, or very few people, if you count extreme cases like George Price, are willing to sacrifice their well-being for others. And many politicians are afraid to sacrifice capital by threatening to raise taxes on the wealthy—whom they rely on to fund their campaigns. "What parties propose is influenced by the views of the rich and by organized business, and pretty much unresponsive to the views of everyone else," Daniel Nettle wrote in 2023.[43] This echoes a further point that Devi Sridhar made to me: being a politician is great if you want to make your friends rich, but awful if you want to make structural changes in society. "This makes you not want to be in public life," she said, speaking of her own experiences as a writer and public health adviser during the Covid-19 pandemic. "The media don't want you there and social media amplifies it."

Improving inequality therefore relies not only on appealing to the needs of the majority, but sacrificing capital to ensure those needs are met. That may be resource capital directly, or social capital—particularly the political variety, which leaders rely on to win elections. Yet fixing major problems like inequality and all its unpleasant descendants requires taking risks, and being willing to pay the costs for improving most people's lives.

Scientists and policymakers in the public health sphere face similar problems. When countries including Austria, France, the United Kingdom, and the United States introduced partial vaccine mandates—at least some people, such as healthcare workers, were required to vaccinate against Covid-19—public trust in science dropped. Those behind the mandates lost political capital, though they didn't sacrifice it: the loss of trust was not an intended outcome of the mandates. Yet in other countries, notably some provinces of Canada and Nepal, mandates did not result in lower public trust. In these cases, officials consulted with members of the public, asked about their concerns, and explained the public health benefits widespread vaccination offered. Understanding the needs of these people aided not only the goal of increasing vaccine uptake, but avoided the pitfalls of the top-down approaches of other countries. And showing an interest in the public need for frank talk and solid information constituted a willingness to pay costs on the part of the government. It requires time and resources to engage with communities effectively.[44]

Sometimes costs are symbolic, and an offer to sacrifice resource capital can have the opposite effect to what is intended. At the time of writing, a series of tragedies is unfolding in a war between Israel and Hamas terrorists in the Gaza Strip. The conflict has been ongoing with some pauses roughly from 2007 onward without any sign of a resolution.

There have been attempts at talks between the Palestinian people and the Israeli government over the last several decades, though no agreement has ever been reached. The anthropologist Scott Atran suggests, however, that offers of land or resources may, surprisingly, undermine attempts to make peace between these people. Instead, as results from work by several separate psychologists show, offers to add economic assistance to deals that may result in a two-state solution can alienate both Palestinians and Israelis—and representatives from the former group said economic packages made them more likely to use violence to achieve their aims.[45]

Symbolic offerings were considered by both groups a more important element of any compromise. Atran notes that Palestinians who participated in his research were willing to openly accept Israel's existence if the Israeli people apologized for their treatment of Palestinians during the 1948 war. Israelis were also more open to agreements between the groups in scenarios where Palestinians recognized Israel's existence.

When in 2007 Atran and his colleagues asked the Israeli prime minister, Benjamin Netanyahu, if he would accept a two-state solution, assuming that Palestinian officials recognized Israel's existence, he responded: "OK, but the Palestinians would have to show that they sincerely mean it, [and] change their anti-Semitic textbooks."[46]

Sincerity, in this case, is demonstrated by a willingness to disavow teachings about anti-Semitism. Yet like trustworthiness, sincerity is something that some people may try to fake. This is critical for diplomacy—but is also a serious problem in nebulous fields like sustainability, where lack of regulation and understanding around interventions like carbon offsetting allows for surprisingly easy exploitation.

Offsetting—or the practice of removing carbon from the atmosphere to compensate for its addition elsewhere—was introduced in the 1990s to create economic incentives promoting sustainable business practices. If a company releases carbon into the atmosphere through running its services or manufacturing processes, it can offset those emissions by, say, planting a number of trees elsewhere that would, theoretically, remove the same amount of carbon from the environment. In exchange for these ostensibly sustainable practices, governments issue carbon credits, which serve as economic incentives for offsetting.

The government-issued credits allow for gains in resource capital—but there are other motivations, too. Many companies that practice carbon offsetting advertise their sustainable practices to gain social capital: they improve their reputations and gain consumer confidence. Carbon neutrality is an attractive advertising aim for a company that generates a lot of carbon, both for economic and reputational reasons.

In practice, of course, the world of offsetting is murkier. Although regulatory frameworks, such as the Gold Standard, do exist, some companies develop their own mathematical calculations of their carbon emissions, and similarly calculate whether their offsetting practices will, in effect, offset those emissions.[47] Companies may therefore make calculations—develop statistical models—that enhance their economic and reputational story most effectively. And their advertisements to this effect—calling themselves sustainable businesses and thereby promoting their products—may misrepresent the good they are, or are not, doing for the environment. This is an example of what's called greenwashing, and there are many instances of it across the energy, beauty, household, and fashion industries.

Gucci, the Italian designer fashion brand, for example, advertised in 2019 that its operations were entirely carbon neutral. In 2023, however, the *Guardian* revealed that the standards of Verra, the company that certified the rainforest carbon offsets on which Gucci (and others) relied for their claims, were insufficient for achieving significant carbon reduction.[48] More than 90 percent of Verra's rainforest carbon credits were "phantoms," the exposé noted. Gucci subsequently deleted claims about carbon neutrality from its website, and ended its partnership with the Swiss carbon credit consultancy it had been working with.

Yet while consumers are, according to an unpublished report from KPMG UK and YouGov, aware of greenwashing in the general sense, people are rarely acquainted with specific examples—possibly because the discussions are so technical that it's difficult to keep up with whether representations of carbon neutrality are genuine.[49] This further damages trust, and is reminiscent of the writer Jon Ronson's warning that to "get away with malevolent power, be boring."[50]

The Cave

These examples—and there are many, many more—point to the fundamental issue in society that we often can't tell who is trustworthy or sincere from who is not. This, coupled with the innumerable effective ways of hiding malevolent intent, can lead to greater inequality, more violence, ineffective public health campaigns, and a collective inability to stop the rapid warming of the planet.

There is, however, something we can do. Although no new idea or technology can solve these problems, despite what many people will tell you, we can try to teach people how, and more important, why, cooperation at the local and global levels is essential for solving each of the problems enumerated here. That requires investment from governments at all levels in the future, in teaching people how to tell what is true from what isn't—and especially about the darkness embedded in human nature.

The Greek philosopher Plato made what seems to me an analogous point in his allegory of the Cave. In this story, people sit in a dark space, ignorant of the world outside their immediate surroundings. They see only the shadows on the cave's wall. These prisoners are then released to the outside world, and learn that the shadows they saw were only partial representations of the truth. The Sun outside the cave taught them about reality: the Sun was the symbol of the form of "the Good," which is the force that, for Plato, guides the universe.

Lenin had in mind technology when he talked of electrification, but—at least in my view—Zoshchenko was thinking of Plato's Cave. The light isn't electrical light or some new, fanciful technology spreading across the world, but rather the truth. And the truth is hard to live with, because we are flawed animals capable of deception, cruelty, and selfishness. Yet confronting that through honest reflection about our evolutionary past gives us the tools to teach ourselves and others about how we can improve the future.

Humans are not just norm followers who exploit others when possible, but norm breakers: we can use our understanding to develop strategies to combat the problems we see. We can be Albert Goering, who breaks rules for good ends; not just his brother, who follows them for ill. "The truth may not make us free, but it should improve our efficiency," Gordon Tullock wrote.[51] Electrification is the spark of insight from which effective interventions must follow.

7
We

There were two in paradise and the choice was offered to them: happiness without freedom, or freedom without happiness. No other choice. Tertium non datur [There is no third possibility].

—Yevgeny Zamyatin, We

Exploitation is everywhere. Plants and animals deceive other organisms about their hidden qualities. Cancer cells trick the immune system into treating them as harmless to the host. Non-human primates, such as chimpanzees, falsify alarm calls to gain access to food.

Humans have something else, which is unseen in the natural world and is unique, beneficial, and dangerous: we have language. We can influence others, hide our identities, and mask our intentions—all to increase our embodied, social, or resource capital. We are all, to a greater or lesser degree, invisible rivals. This is as core to being human as is what Darwin called the social instinct. Yet unlike with any other species, it is our choice whether we try to use people or to work with them—

the fundamental opposition that grounds Pyotr Kropotkin's view of the world against Thomas Hobbes's *Leviathan*.

These incompatible views of human nature are, however, just stories. And a popular story of today mirrors Kropotkin's: that we have eliminated cheaters from our social groups. Overt, dominant alphas, the thinking goes, were killed in our distant past; norm-breakers were punished; non-contributors were cast out. Today, our societies are mostly internally cooperative, and the problems we now see arise solely because of competition between groups of people, rather than because of people within groups.

This makes for a good bedtime story that might help many people get to sleep—but it is a lie. It misses out on a key element of human nature that only introspection and anthropology can reveal: the ability to deceive ourselves and others, to hide while waiting for an opportunity to exploit. We compete at every level, and when we can't win, we often cheat.

Cancer is the story of how one cell in the body can betray its host, inevitably killing it in the process. Invisible rivalry is the same story—just applied to societies, rather than organisms. Consequently, in contrast with what many people believe today, we did not solve the free-rider problem. And just as clinical interventions and the body's natural immune system force cancer to change strategies for growth, our own social interventions forced free riders into hiding. We traded Macbeth for Iago, Pompey for Caesar—and the result, in the modern world, is the success of despotic, Machiavellian figures like Vladimir Putin and Donald Trump.

Invisible rivalry is the social equivalent of cancer. It is the selfish core against which all human cultural diversity is fortified, though as often is true with cancer, victory is not possible. The choice, then, isn't between Kropotkin and Hobbes, but between accepting our own flawed heritage, and working toward a cooperative future—or telling ourselves false stories and allowing invisible rivalry to overwhelm us.

Accepting the problem—shining a light in our collective cave—is just the first step. Expecting others to behave cooperatively in critical spheres like climate change, inequality, and technological regulation is the more difficult task. What would you say to a cancer cell if you could communicate with it? You might explain to it that its endless reproduction is going to end up killing the body that hosts it—and thereafter, all the cell's descendants. The cell might agree its exploitation of the cooperative cells around it is shortsighted, but might keep it up anyway.

Humans, unlike cells, don't have to be like this. We don't, unlike much of the non-human world, rely entirely on our genetic programming to guide our behavior. We can influence others to work together, build mutual trust, and change our societies for the better. The free-rider syndrome, manifested as invisible rivalry, doesn't need to break us.

The view of human nature we accept determines whether and how we trust others, whether at the individual, community, or structural level. If we accept the view that humans are fundamentally prosocial cooperators, we will be more likely to trust blindly. If, conversely, we believe everyone is selfish, we won't trust anyone.

Both extremes are likely to result in a breakdown in the norms that govern society. Blind trust results in easy exploitation by bad actors across society; little or no trust allows for the easy demonization of others. Where there is no trust, following the arguments of researchers like Gordon Tullock and Nicholas Humphrey, despotism is likely to be the result. The social dilemma, here, is easy to exploit: manipulative leaders who insist that the public's security is at risk are likely to be successful at attracting support. It is simple to say that some specific groups, such as Jewish people, Islamic people, Republicans, Democrats, the far left, the far right, White Christians, Zionists, transgender people, trans-exclusionary radical feminists, big corporations, or any other category of humans, are collectively responsible for our problems. Autocrats and those intent on being autocrats are effective at using this tactic—and then stifling dissent against their narrative once they take power.

Thereafter, trust is irrelevant—whether there is too much or too little of it. Where there are no personal freedoms, it doesn't matter whether you trust your government, as the circumstances of people in Yevgeny Zamyatin's *We*, George Orwell's *1984*, and Kurt Vonnegut's *Harrison Bergeron* show. Much as the philosopher Ludwig Wittgenstein argued that doubt has sense only where knowledge is possible, so too is trust only relevant where there is freedom.[1] Autocracy, the basin of attraction toward which open societies are drawn, eliminates the need for trust altogether.

How to Place Trust

The optimal state is therefore neither too much nor too little trust—and more specifically, one where the public understands how to place trust discerningly. One analysis of previous studies in this huge body of literature found that in a lab setting, people are only just better than chance at lie detection—about 55 percent, with little difference at the individual level.[2]

The strongest factor that predicted accuracy was, it turned out, the perceived credibility of the source. When a liar or a truth-teller appeared credible, the responding participant was more likely to rate what was said as honest. Some people just appear more trustworthy—and it's those people that we tend to believe, regardless of whether they are being honest.

Trends on social media, which increasingly drive political sentiment across the world, mirror these findings. The blue tick mark, as one example, was developed by companies like Twitter to, at least in part, separate trustworthy information sources from others. An account called the *Financial Times* with a blue tick was more likely to have true information than an account with the same name without one. Since the tech billionaire Elon Musk's takeover of the platform, however, anyone can pay for a blue tick—making a signal previously associated with credibility an accessible commodity. In mid-2023, just after this change was introduced, fake accounts impersonating major figures, including Amazon's founder Jeff Bezos and the late U.S. senator John McCain, appeared with blue ticks.

What's more, and I think more worrying: we reward liars. People tell us when we are young that cheaters never prosper, but they mean unsuccessful cheaters lose out. The liars—cheaters in the game of information exchange—do prosper, at least when they are not detected. These deceivers, some psychologists say, are reinforced in their behaviors through success. When lying brings benefits early on in life, people will continue to lie, while those who are caught will stop. "The deceptively rich get richer, and the poor stay the same," the authors of the lie detection analysis wrote.[3]

We do, in other words, what works for us. This is true whether we're talking about a little boy who lies about brushing his teeth before bed or a multinational conglomerate that misrepresents its sustainability initiatives. In both cases, success and defeat reinforce behaviors positively or negatively. And this bias in our psyches can lead us down dangerous paths, both to ourselves and society.

Again, social media platforms like X illustrate how. Writers including Joyce Carol Oates and Jon Ronson have argued that the ways others reinforce our views on social media can drive us to make more and more extreme claims. Someone who develops an influential social media profile may initially make reasonable points about controversial topics, whether about rights for transgender people, warfare in the Middle East, or the importance of wearing masks to prevent Covid-19 transmission. Yet over time, as that person becomes aligned with a particular group online, the posts may become less reasonable, or even toxic—driven by a desire for likes, reposts, and followers, which are the digital world's modes of reinforcement.

This is one example in the broader sphere of how we seek to optimize resource, social, and embodied capital in our interactions with others—and how that drive can negatively affect society. In the digital sphere the effects of reinforcing falsity and extremism may be limited. Yet when reinforcement in the real world affects how politicians behave, as my colleague Devi Sridhar has said, we can start to see problems both in our political systems and in the kinds of people that vie for power.

Prosociality, in other words, is not always—or even often—rewarded in human cultures. It is better to appear trustworthy and to be selfish, and to exploit others when you have the opportunity to do so. It is better, evolutionarily speaking, to be an invisible rival.

At the individual level, there are two twin solutions. The first is self-knowledge—understanding that we all have the capacity to be invisible rivals, and as a consequence we each have a moral choice about whether and how to exercise this power. The second is understanding others: learning to look for signs that indicate when someone is trying to exploit your beliefs or behaviors.[4]

The best any of us can do, for both, is to arm ourselves intellectually with the power to tell who is credible and who is not—and to apply that framework to our own actions. This is a dull task that requires close analysis, both of our own beliefs and the understanding of others' motivations—but in a world increasingly distorted by falsity, it is the only tool we have. Yet through education, questioning, introspection, and listening carefully, it is possible.

Think, for one thing, about whether and how people take in information about current affairs. Some people who write about the intrinsic good of human nature argue that avoiding the news is a good thing. And while I agree that too much news—especially, it seems, in recent years—is likely to be bad for your mental health, I can't imagine worse advice for anyone whose aim is to help improve society. Ignoring local and global affairs is exactly what bad actors across the world want people to do: it's much easier to exploit those who don't know they are being exploited.

To do this, it's critical to engage with media enough to understand how vying social and political interests create the problems we see across societies today. Doing so requires time, effort, and the understanding that everyone who speaks, in whatever medium, has an agenda. And yet teasing that agenda out in a way that allows us to place trust effectively in sources is critical.

Our most important tool for doing this is education, both of ethics and critical thought. Neither needs to be formal—and often cannot be, such as when fiscally conservative governments defund educational programs to a breaking point, or where a party prefers an uneducated populace (a view often associated with the Republican party in the United States).[5]

Instead, education can involve parents and communities teaching younger people, encouragement to read, or even just religious instruction around ethical thinking. The point is to teach people to think ethically for themselves, and to give them the tools to do so. The subjects, whether math, religion, language, philosophy, life science, engineering, and so on, matter less than the raw ability to question effectively. (And if you need more motivation than that to keep up with education, research by my friend and colleague Carol Brayne and associates revealed that the longer people stay in education, the less likely they are to develop dementia.[6] It is good for you to learn to listen, not only to better understand your world, but probably also for your health.)

Furthermore, given our increasingly complex media and social media environments, as well as the pervasiveness of disinformation, the individual ability to analyze information is critical. Some researchers have developed tools for "inoculating" people against disinformation—but for me, the critical intervention, as with the policy level, is understanding.[7] Even reading a short and simple introduction to a science like statistics or a humanities subject like history can be sufficient to alert you to when someone is trying to hide something.[8] Often, what is hidden in a communicative interaction is more important than what is said, and giving people the tools to discern that hidden information is the key to understanding a speaker's intention.

A good understanding, moreover, motivates asking important questions. Some time ago, I met a philosopher interested in the origins of questions. She told me that while people often assume that the person speaking in declaratives holds the power in a conversation, she believes instead that power rests with those who inquire. Those who ask questions can drive the flow of information. The answers they receive may not be complete or honest, but interrogating an idea or way of thinking is an essential part of discourse in any society.

There is, of course, such a thing as a stupid question—("are humans cooperative or competitive?")—but being unafraid to ask them, and I have asked a lot of stupid questions over my career, is a critical part of learning to listen effectively. One of these, and perhaps the most important in conversations about exploitation in society, is, Who benefits? When someone tells you something novel in everyday dialogue, in a seminar, on an online forum or a social media platform, and so on, the essential question should always be: What does this person gain from my believing this statement?

Beliefs about motivations drive how people perceive major ongoing problems, such as climate change. A poll in 2016 found that only about half of adults in the United States say that global warming is caused by human activities. Other research suggests that these perspectives are driven by the belief that climate scientists are incentivized by advancement in their careers or their own political views. And to the latter point, there is evidence of what's called white hat bias in the sciences: when researchers believe they are on the side of what is moral, they are less likely to be impartial when coming to conclusions based on their data. Trust, in other words, is damaged when listeners—those receiving information from scientists—believe communicators have ulterior motives.

Scientists, like everyone else, have good and bad motivations for promoting their work. We should question those motivations. But we should also question the motivations of people telling us that climate change is not real, or that people aren't causing it. There is evidence that executives at the oil and gas company now known as ExxonMobil were aware nearly fifty years ago that climate change is a real, human-driven phenomenon.[9] Still, the company invested heavily in conservative think tanks that helped to create doubt. Are their motivations better or worse than those of scientists trying to attract funding and boost their careers? Find out and decide.

Our proclivity for self-deception, however, makes questioning motivations especially difficult.[10] Can (and should) we question the trustworthiness of others when we can trick ourselves? Many people invested in them have good motivations for their arguments, and many others seek to free ride on whatever is the most popular view—and gain social capital in consequence. It's difficult, if not sometimes impossible, to tell the difference between a devout believer in a principle compared with someone who fakes belief for social benefit, but when the latter are exposed, the movement, whatever it is, can be damaged.

Of course, some people have multiple motivations for expressing their views, and risk to social capital is rarely something anyone can ignore. But that doesn't mean we can't use tactics to unpack our own motivations—and in doing so, lift the layers of self-deceit that can prevent us from understanding why we hold some of the views, or say some of the things, we do.

The late philosopher Bernard Williams offered a helpful test. Imagine, he suggested, that you are a doctor treating a patient for a life-threatening condition.[11] You can then choose to live in one of two worlds: in the first, you successfully treat the patient, who recovers, while you spend the rest of your life thinking the patient has died. In the second, the patient dies, and you spend your life thinking you saved that person. Which world would you rather live in?

Williams's idea is broader than the world of clinical medicine, and gives each of us the tools necessary for unmasking the motivations and biases we find inconvenient. Would you rather live in a world where climate change was solved and you had nothing to do with it, or to receive the adulation for solving climate change while the real world burns? Would you rather receive acclaim for your veganism while a thousand other people kill and eat animals, or receive no acclaim and for the same thousand people to become vegan?

Everyone, even the most manipulative people among us, would benefit from planning for the long term. Jon Ronson told me, following his own conversations with a well-known psychologist, that convincing psychopaths that their own long-term interests are more important, for their own well-being, than are short-term desires is the only way to get them to change their behaviors. Yet we could all benefit by following the same advice, much as a cancer cell could. Interrogate your beliefs, unpack your biases, and always ask yourself whether you are motivated in your views by what you want others to think of you, or what kind of world you want to live in.

How to Deserve Trust

Trust is, of course, a two- (or more) sided relationship, and it falls on communities, institutions, and political bodies to demonstrate that they are worthy of trust, regardless of whether the population they serve is discerning.[12] The possibility of faking trustworthiness, however, remains a problem: anything honest people say to win over support can be copied by a deceiver.

Luckily, research in specific circumstances—such as ordering a cab in New York or trying to make in the Sicilian mafia—reveals when and how people can effectively display their trustworthiness relative to their in-group. Diego Gambetta, a sociologist who has explored the problem of demonstrating trustworthiness for the past forty years, has shown that where people's lives or capital are at risk, they become adept at detecting signals that suggest a person can be trusted. Gambetta's book *Codes of the Underworld*, published in 2009, shows that among criminal gangs, people learn to pay costs that earn them trust in their groups, leading to a complex semiotic system in which it is difficult to misrepresent yourself—at least among fellow criminals.[13]

There is no reason that this work—linking signaling theory to the problem of demonstrating trustworthiness—cannot apply more broadly to modern institutions. And one simple way that political leaders can show they deserve trust is through listening to the people they govern. Gillian Tett, an anthropologist and writer, argues in *Anthro-vision* that listening—being someone who learns to understand people—is critical for creating any solutions for social issues.[14] This is true in businesses as they expand to new areas, and for policymakers as they attempt to introduce public health rules such as vaccine mandates and sustainable practices. You cannot address an issue that you do not understand, as Elinor Ostrom's work on local social norms indicates.

Given that climate change is a global issue, and one that moreover predominantly affects the developing world, speaking to local communities and indigenous populations is essential in aims to address it.[15] First Nations people are collaborating with local authorities in North America to co-create strategies for combating the growing burden of forest fires. This has already had some success in California, and the number of partnerships of this kind is growing.[16]

Similarly, Inuit people in eastern Canada partnered with researchers on the Igliniit Project, which helped both researchers and indigenous people to map out the effects of climate change in the territories they live in.[17] The researchers involved mounted a GPS receiver, weather tools, and a digital camera on the local hunters' snowmobiles. As they traveled, the tools measured weather data such as temperature, and also allowed the users to update local maps with information about the changing environment. This allowed researchers to track, year over year, how climate change was affecting the local area.

Although the world is rapidly changing and it is not possible to fully control climate change or feed growing human populations without affecting the environment, we will be paralyzed to introduce any interventions without local knowledge.

Indigenous people make up about 6 percent of all humans worldwide—and to the extent that governments displace them or disrupt their ancient ways of life, the loss of knowledge will have catastrophic environmental effects.

In August 2023, the United Nations Institute for Training and Research hosted an event that aimed to address this problem: the first ever Global Indigenous Youth Summit on Climate Change. This twenty-four-hour global dialogue aimed to bring young indigenous people together to discuss local approaches for addressing climate change. One major warning from this event was the view that Western powers, by exploiting environments for materials such as nickel, lithium, cobalt, and copper, are causing huge amounts of damage. Although these materials are, ironically, important for future technologies that may power relatively sustainable technologies such as electric car batteries, ignoring the local knowledge of indigenous people is an unsustainable strategy.

The next generation of researchers and policymakers across the health, economic, and climate sciences must then also be anthropologists: we should shut up and listen. We need to listen to people from poorer areas across developed and developing countries and to people living in slums that we insist should be converted to—often mediocre—social housing. We need to listen to people with mental health problems who, research shows, are nearly twice as likely as others to develop serious physical ailments.[18] The voices of the marginalized are the voices necessary for creating the egalitarian world we pretend we've lost. We didn't lose it, because we never had it, and we never had it because there is always a group of people—women, young men, religious minorities, and so many others—society marginalizes. The mode of exploitation is always clearest to those being exploited.

Yet despite our collective shortcomings, there are ways we

can help ourselves and others to become good listeners. What each of us needs is just an open but critical mind and a motivation to expose modes of exploitation wherever they are—even if we are the source. Amartya Sen, the economist, philosopher, and Nobel laureate, writes that to create a fairer world, we don't need to develop some ideal form of justice. Instead, we need to identify injustices. And to do so—at least, in my view—we must start with the injustices we, individually, perpetuate.[19]

How to Treat the Free-Rider Syndrome

Even in the most educated, discerning society, invisible rivalry will persist—it is the incurable syndrome through which exploitation will always arise, threatening the social norms that hold cultures together. Fostering trust, then, depends not only on having an educated populace, but on structural interventions that help to prevent exploitation.

Here again, the world of oncology can give some guidance. Despite the many advantages cancer has against us, oncologists are increasingly using evolutionary theory to predict how the disease will change, and to use those predictions to prevent the disease from spreading. If you can use theory and mathematics to predict how the disease is going to change, you can stay one step ahead of it.

One of these approaches is called adaptive therapy, and relies not on destroying a tumor entirely, but in manipulating the cancer's cellular makeup to prevent it from spreading. This alternative strategy can lead people to live longer with their disease, and possibly without the side effects accompanying aggressive treatments like chemotherapy. The goal with this strategy is not to eliminate cancer, but to enable people to live without it growing out of control.

Researchers and policymakers working to prevent free

riding in the modern world—whether that is tax evasion, benefits fraud, or any other cheating social behavior—can take an analogous approach. We can't eliminate free riding, but we can learn to live better with it.

For example, we can accept, relying on data from experimental economics and psychology, that roughly 30 percent of people are likely to behave selfishly in social interactions. A subset of these free riders may aim to cheat social programs for personal benefit. Yet that cheating should not invalidate our support systems: we need, instead, to model cheating into our frameworks, rather than use free riding as an excuse to end welfare spending.

This is also true at the level of global efforts to slow climate change. Nearly two hundred entities signed the Paris Agreement in 2015, which was a global effort to ensure global temperatures remain less than two degrees Celsius warmer than those during the industrial era. But many of the signatories, including governments, are not on track to keep their promises. Although there was a 5.2 percent decrease in global fossil carbon emissions at the outset of the Covid-19 pandemic, 2022 set a new record with an estimated 37.5 billion tons, according to data from the Global Carbon Project.[20]

One study found that several countries—Russia, Iran, and Saudi Arabia—are "critically insufficient" in meeting their climate pledges and their work toward the goal of the Paris Agreement. Australia, Brazil, Canada, China, and India were found by the same analysis to be "highly insufficient."[21]

The temptation to ignore these pledges is strong. Countries involved in economic conflicts, such as the United States and China, are motivated to place financial interests ahead of climate issues. Others, such as Russia, are engaged in military conflict, and consequently place their ambitions for conquest—

and maintaining a faltering economy by relying on natural gas, in this example—above climate goals.

Given these shortcomings, and short-sighted thinking, in other governments, it may be difficult for other countries not to sacrifice their own pledges to help boost faltering economies. This is especially true in democracies with increasingly erratic political parties: arguments about failing economies can sway elections, sometimes in the direction of party leaders seeking to become autocrats.

Again, however, the presence of free riders should not dissuade countries from pursuing their climate agendas. Part of that is analogous to accepting individual cheaters in welfare systems: some people, and some groups, are just going to cheat, no matter what they say publicly. Yet that does not mean the efforts of others are likely to be fruitless, or that governments that do not follow through on their pledges will even succeed economically. The Inflation Reduction Act of 2023 in the United States aimed to both boost sustainability innovation and attract economic investment—and through late 2024, it appeared to be succeeding at both, despite an initial cost of $500 billion.

The mistake of models created from the Paris Agreement is to assume that all countries will abide by their pledges. They will not. And adjusting for that can help us make more realistic calculations about how effectively we are combating climate change. This does not require singling any country out as dishonest, but rather assuming that a portion of the reduction in carbon emissions—probably, again, around 30 percent—will not happen. More realistic expectations help us to maintain momentum, and not be lost in short-term squabbles.

Focus on punishment in any of these systems—broadly in the category of social dilemmas, where people, institutions, or governments benefit from free riding—is unlikely to be effective. For whatever reason, people in many cultures feel motivated to punish free riders, sometimes at great cost to themselves. Yet exposure may be a stronger force: it's better to be adaptive—to show that a person or group is free riding on a system—than to try to punish them. Loss of social capital through reputation is an important motivator for anyone, and journalistic work exposing exploitation can be as effective at driving behavior change as any putative legal penalty.

In the sustainability sphere, this may be especially important, given the number of complex and boring ways organizations and governments may hide their non-sustainable practices. Focusing on your own practices, and helping your organization or government to focus on theirs, is a more effective use of time. Ignore spiteful impulses, expose exploitation where possible, and move on.

To solve the problem invisible rivalry poses, then, we need to accept it—much as one way it may be possible to prevent people dying from cancer is to enable them to live with it.

This leads to a broader view of an expanding circle that we can adopt to solve the free-rider problem. First, and most important, we must admit there is a problem, and ensure that people, including ourselves, understand that exploiting society for self-benefit is wrong. Teaching ethics—and understanding why ethical behavior is essential for an enduring society—is the center.

Second, parents, communities, and society more broadly should educate people about how to place trust discerningly, and to avoid those who aim to exploit them. And third—and only as a final resort—should we focus on punishing exploiters. Helping people to understand why living ethically is important is preferable to attempting to force their behaviors through enforcement.[22]

Although some of my colleagues promote punishment

as the most effective way to control free riding, and suggest we work punishments into modern life at the community level, this, in my view, leads to a world that resembles Zamyatin's *We*, which shares some themes with George Orwell's better known and later published work, *1984*. An open society of discerning, educated people must be better than a closed one that controls behavior through punishment and coercion. We can't overcome the problem of opportunity; we can only help others understand that to be an opportunist is harmful, in the long term, for us all.

Depressing Bedtime Stories

Arthur Machen, the nineteenth-century author of several famous supernatural stories, wrote in *The Three Impostors* that many innocuous characters in stories of fantasy—such as fairies— have darker origins than we might like. The image of diminutive, benevolent creatures with wings that J.R.R. Tolkien also detested represent a collective whitewashing of a horror we don't want to face.

To me, complex explanations for how and why we are fundamentally cooperative serve the same function. It is simpler and easier to say that people are benevolent and that extrinsic forces or rare bad actors cause our problems. It is harder and more depressing to say that each of us has these qualities, to a greater or lesser degree, and that we choose whether to exercise them. The reality of living in a world with finite resources is not a pleasant one—but facing the truth underlying rosy stories gives us the tools to protect ourselves. If a system can be exploited, it will be.

The dilemma each of us faces now is whether to confront invisible rivalry or to let exploiters undermine society until the democracy in the free world unravels—and the freedom of dissent is gone. This is a real and pressing problem, evidenced by the rise of the Julius Caesar of our time—Donald Trump—but one which the story of evolution has predicted since the origins of life and later, language, and which will only change form again even if the current crises are overcome. Hearing the less pleasant story, and perhaps more important, facing your own selfish core, is not easier—but it will make our lives, and the lives of those in the future, better. As Hope Mirrlees wrote in *Lud-in-the-Mist*:

> You should regard each meeting with a friend as a sitting he is unwillingly giving you for a portrait—a portrait that, probably, when you or he die, will still be unfinished. And, though this is an absorbing pursuit, nevertheless, the painters are apt to end pessimists. For however handsome and merry may be the face, however rich may be the background, in the first rough sketch of each portrait, yet with every added stroke of the brush, with every tiny readjustment of the 'values,' with every modification of the chiaroscuro, the eyes looking out at you grow more disquieting. And, finally, it is your own face that you are staring at in terror, as in a mirror by candle-light, when all the house is still.[23]

Glossary

Below is a list with definitions of important terms central to discussions in this book. Most are adapted from Stuart A. West, A. S. Griffin, and A. Gardner, "Social Semantics: Altruism, Cooperation, Mutualism, Strong Reciprocity, and Group Selection" (2007), except where indicated otherwise.

(Biological) Altruism: a behavior in which an agent pays a cost to grant a benefit to a recipient, both understood in terms of direct fitness only; this does not include inclusive fitness effects. This is understood in terms of effects rather than motivations.

(Psychological) Altruism: a behavior intended to help another person or animal without ulterior motive. This is understood in terms of motivations rather than effects. (From Richard Kraut, "Altruism," *The Stanford Encyclopedia of Philosophy*, Fall 2020 edition, edited by Edward N. Zalta, https://plato.stanford.edu/entries/altruism/.) Based on arguments in chapter 4, I define non-biological altruism as the demonstrable motivation to sacrifice capital for another.

Cheating: lack of or insufficient cooperation while taking benefits from the cooperation of others; used interchangeably with "free riding."

(Biological) Cooperation: a behavior through which an agent gives a benefit to a recipient; the behavior is selected for because of its effect on the recipient, which usually results in a consequent benefit for the agent.

Direct fitness: the effects of an agent's behaviors on the number of surviving offspring.

Inclusive fitness: the effects of an agent's behaviors on the number of surviving offspring and the number of surviving offspring of the agent's genetic relatives.

Relatedness: measurement of genetic similarity, relative to the overall population in question.

Self-interest: behaviors, whether intended or otherwise, that promote an actor's biological or social interests in maximizing capital; this may involve selfishness, cooperation, both, or neither. This is my own definition from reading the relevant literature.

(Biological) Selfishness: a behavior that is beneficial to the agent and imposes a cost on the recipient; understood only in terms of effects.

(Psychological) Selfishness: a behavior intended to promote one's own interests without regard for the effects on others. This is my own definition from reading the relevant literature.

Notes

1
A Hero of Our Time

1. Many contemporary academics and writers, including Joseph Henrich, Martin Nowak, David Sloan Wilson, Nichola Raihani, and Rutger Bregman, are proponents of this view.

2. Edward Helmore, "Dan Price Resigns as CEO of Payments Firm After Misconduct Allegations," *The Guardian*, August 18, 2023.

3. Helmore, "Dan Price Resigns as CEO of Payments Firm After Misconduct Allegations."

4. This term was notably used in a paper in *Nature* by Samuel Bowles and Herbert Gintis in 2002: Bowles and Gintis, "Homo Reciprocans."

5. For key terms such as free rider, cooperation, and altruism, academically accepted definitions appear in the Glossary.

6. Clark et al., "Prosocial Motives Underlie Scientific Censorship by Scientists."

7. Jon Ronson, "How One Stupid Tweet Blew Up Justine Sacco's Life," *New York Times Magazine*, February 12, 2015, https://www.nytimes.com/2015/02/15/magazine/how-one-stupid-tweet-ruined-justine-saccos-life.html.

8. Ali Vingiano, "This Is How a Woman's Offensive Tweet Became the World's Top Story," *BuzzFeed News*, December 22, 2013, https://www.buzzfeednews.com/article/alisonvingiano/this-is-how-a-womans-offensive-tweet-became-the-worlds-top-s.

9. Taken from a letter from Swift to Alexander Pope; Kathleen Williams, introduction in Swift, *Tale of a Tub*.

2
Smoke

1. Joe Sommerlad, "Robert Hendy-Freegard: The Real Story of The Puppet Master Conman," *Independent*, February 3, 2022, https://www.independent.co.uk/arts-entertainment/tv/news/robert-hendy-freegard-puppet-master-mi5-b2006686.html.
2. Southern Poverty Law Center, "Changing of the Guard."
3. Stein, "How Cult Leaders Brainwash Followers for Total Control," *Aeon*, June 20, 2017; Malcolm Gladwell, "Sacred and Profane," *New Yorker*, March 24, 2014, https://www.newyorker.com/magazine/2014/03/31/sacred-and-profane-4; Mendy Levy, as told to Jacob Shamsian, "I Fled an Extremist Jewish Cult in Guatemala When I Was 15 Years Old. I Grew up with Virtually No Education and Wasn't Allowed to Show Love to My Parents," *Business Insider*, November 7, 2021, https://www.insider.com/lev-tahor-jewish-extremist-group-life-after-escape-2021-11.
4. Stein, "How Cult Leaders Brainwash Followers for Total Control."
5. Levy, "I Fled an Extremist Jewish Cult in Guatemala When I Was 15 Years Old."
6. A more technical definition for "cooperation" in biology is "a behaviour which provides a benefit to another individual (recipient), and which is selected for because of its beneficial effect on the recipient"; West, Griffin, and Gardner, "Social Semantics."
7. Montaigne, *Essays*.
8. Rousseau, *Discourse on the Origin and Foundations of Inequality Among Men*.
9. Kropotkin, *Mutual Aid*.
10. Kropotkin, *Mutual Aid*.
11. He writes in *Mutual Aid*: "The readiness of the Russian zoologists to accept Kessler's views seems quite natural, because nearly all of them have had opportunities of studying the animal world in the wide uninhabited regions of Northern Asia and East Russia; and it is impossible to study like regions without being brought to the same ideas."
12. This is portrayed in Cyril Schäublin's film *Unrest* (or *Unrueh*), from 2022.
13. Wrangham, "Targeted Conspiratorial Killing, Human Self-Domestication, and the Evolution of Groupishness."
14. Lonsdorf, "Sex Differences in the Development of Termite-Fishing Skills in the Wild Chimpanzees."
15. Harmand et al., "3.3-Million-Year-Old Stone Tools from Lomekwi 3, West Turkana, Kenya."

16. Phillips, Li, and Kendall, "The Effects of Extra-Somatic Weapons on the Evolution of Human Cooperation Towards Non-Kin."
17. Stibbard-Hawkes, *Egalitarianism and Democratized Access to Lethal Weaponry.*
18. Hill and Hurtado, "Cooperative Breeding in South American Hunter-Gatherers."
19. Hill, "Prestige and Reproductive Success in Man."
20. Aktipis, Cronk, and Aguiar, "Risk-Pooling and Herd Survival."
21. Levinson, "Introduction."
22. Armstrong, *Rossel Island.*
23. Stibbard-Hawkes, Attenborough, and Marlowe, "A Noisy Signal: To What Extent Are Hadza Hunting Reputations Predictive of Actual Hunting Skills?"
24. Venkataraman, Kraft, and Dominy, "Tree Climbing and Human Evolution."
25. Chaudhary et al., "Competition for Cooperation."
26. Moore, "The Evolution of Exploitation." To date, the articles have fewer than ten citations—nearly equivalent, in academic language, to being ignored; Moore, "The Exploitation of Women in Evolutionary Perspective."
27. Moore, "The Exploitation of Women in Evolutionary Perspective."
28. Khoekhoen is translated to English as "men of men."
29. This is a direct quotation, but "Hottentot" is an outdated and racist term for the Khoekhoen.
30. Holmberg, *Nomads of the Long Bow.*
31. Or, in other words, not egalitarianism.
32. Singh and Glowacki, "Human Social Organization During the Late Pleistocene."
33. Stewart, "Investigating the Calusa."
34. Hannah Moore, "Are All the Ants as Heavy as All the Humans?" *BBC News*, September 21, 2014, https://www.bbc.com/news/magazine-29281253.
35. Schultheiss et al., "The Abundance, Biomass, and Distribution of Ants on Earth."
36. Kaplan et al., "A Theory of Human Life History Evolution."
37. James, "Bergmann's and Allen's Rules."
38. Singh and Henrich, "Why Do Religious Leaders Observe Costly Prohibitions?"
39. Chagnon and Irons, *Evolutionary Biology and Human Social Behavior.*
40. Moore, "The Evolution of Exploitation."
41. More recently, economists Daron Acemoglu and James Robinson used the term "extractive state" for this set of circumstances; Acemoglu and Robinson, *Why Nations Fail.*

42. Singh, "The Idea of Primitive Communism Is as Seductive as It Is Wrong."
43. Lee and DeVore, *Kalahari Hunter-Gatherers*.
44. Humphrey, "The Social Function of Intellect." Though, ironically, the person credited with this idea, Nicholas Humphrey, prefers "social brain," because he doesn't believe all social intelligence is Machiavellian (and neither do I).
45. Chartrand and Bargh, "The Chameleon Effect."

3
The Evolution of Invisible Rivalry

1. Plato, *The Republic*, Book Two.
2. Byrne and Whiten, "Cognitive Evolution in Primates."
3. Mills, "Unusual Suspects."
4. Haridy et al., "Triassic Cancer—Osteosarcoma in a 240-Million-Year-Old Stem-Turtle."
5. Dawkins, *The Blind Watchmaker—Why the Evidence of Evolution Reveals a Universe without Design*.
6. Simpson, *Tempo and Mode in Evolution*.
7. Aktipis, *The Cheating Cell*.
8. Bernard Mandeville, *The Fable of the Bees*, 1732.
9. Smith, *The Wealth of Nations*.
10. Hamilton, "The Genetical Evolution of Social Behaviour"; Axelrod, *The Evolution of Cooperation*; Trivers, "The Evolution of Reciprocal Altruism."
11. Grafen, "Modelling in Behavioural Ecology."
12. Trivers, "The Evolution of Reciprocal Altruism."
13. Axelrod and Hamilton, "The Evolution of Cooperation."
14. Dunbar, "Coevolution of Neocortical Size, Group Size, and Language in Humans."
15. Alexander, *The Biology of Moral Systems*; Nowak and Sigmund, "Evolution of Indirect Reciprocity."
16. Dunbar, *Grooming, Gossip, and the Evolution of Language*.
17. Dunbar, *Grooming, Gossip, and the Evolution of Language*.
18. Garfield et al., "The Content and Structure of Reputation Domains Across Human Societies."
19. Hamilton and Zuk, "Heritable True Fitness and Bright Birds."
20. Dey, Dale, and Quinn, "Manipulating the Appearance of a Badge of Status Causes Changes in True Badge Expression."
21. Zahavi, "Mate Selection—A Selection for a Handicap."
22. Grafen, "Biological Signals as Handicaps."

23. Reby and McComb, "Anatomical Constraints Generate Honesty: Acoustic Cues to Age and Weight in the Roars of Red Deer Stags."
24. Maynard-Smith and Harper, *Animal Signals*.
25. Bond and Depaulo, "Individual Differences in Judging Deception: Accuracy and Bias"; Fonseca and Peters, "Is It Costly to Deceive?"
26. West-Eberhard, "Sexual Selection, Social Competition, and Speciation"; see also Nesse, "Social Selection Is a Powerful Explanation for Prosociality."
27. Noë and Hammerstein, "Biological Markets."
28. Hare, "Survival of the Friendliest."
29. Wrangham notes that hunter-gatherers have lower rates of reactive aggression than our closest genetic relatives, chimpanzees.
30. See, for example, Henrich, *The Secret of Our Success*.
31. Jesse Eisinger, Jeff Ernsthausen, and Paul Kiel, "The Secret IRS Files: Trove of Never-Before-Seen Records Reveal How the Wealthiest Avoid Income Tax," *ProPublica*, June 8, 2021, https://www.propublica.org/article/the-secret-irs-files-trove-of-never-before-seen-records-reveal-how-the-wealthiest-avoid-income-tax.
32. Kevin M. F. Platt, "The Profound Irony of Canceling Everything Russian," *New York Times*, April 22, 2022, https://www.nytimes.com/2022/04/22/opinion/russian-artists-culture-boycotts.html; Jacqui Goddard, "War in Ukraine: Netflix Shelves Tolstoy Adaptation After Criticism," *Times*, March 4, 2022, https://www.thetimes.co.uk/article/war-in-ukraine-netflix-shelves-tolstoy-adaptation-after-criticism-vwwj8pvn3.
33. Wallace, Buil, and de Chernatony, "'Consuming Good' on Social Media."
34. Some people call others that behave like this "overnight activists," and this is a theme that features in the song "family ties" by Baby Keem and Kendrick Lamar.
35. Goodman, "The Problem of Opportunity"; this is the central argument of my own Ph.D. work.
36. Jillian Jordan and David Rand, "Are You 'Virtue Signaling'?" *New York Times*, March 30, 2019, https://www.nytimes.com/2019/03/30/opinion/sunday/virtue-signaling.html.
37. Singh and Hoffman, "Commitment and Impersonation."
38. This is typically viewed as a group-level quality in the evolutionary sciences literature. For example, groups of people that planned effectively could out-compete (in warfare or otherwise) groups without effective planning. Here, instead, I am interested in how proactive aggression, which is thought to be over-expressed in psychopaths, manifests at the individual, not group, levels.
39. Humphrey, "The Social Function of Intellect."

4
Capital

1. Mauss, *The Gift*.
2. This formulation is associated with Christian ethics; the Judaic formulation takes the negative "do not treat others as you would not like to be treated."
3. Parfit, *On What Matters*.
4. Hume, "A Dialogue." "Rhone" and "Rhine" are both Celtic names, and derive from the Indo-European daughter language the Celts spoke when they lived in central Europe until about 700 BCE.
5. Interestingly, and as far as I can tell, unrelatedly, Parfit thinks of the project of improving human ethics as "climbing the mountain."
6. Smith, *The Wealth of Nations*; see also Fisher, "What Is Capital?"
7. Erasmus, *In Praise of Folly*.
8. DeMerchant and Roach, "Vocal Responses of Hermit Thrush (Catharus Guttatus) Males to Territorial Playback of Conspecific Song."
9. Koenig et al., "Natal Dispersal in the Cooperatively Breeding Acorn Woodpecker"; Bowles, Smith, and Borgerhoff Mulder, "The Emergence and Persistence of Inequality in Premodern Societies."
10. Orians, "On the Evolution of Mating Systems in Birds and Mammals."
11. Hilary Weaver, "Scarlett Johansson Doesn't Think Monogamy Is Natural," *Vanity Fair*, February 15, 2017, https://www.vanityfair.com/style/2017/02/scarlett-johansson-marriage-and-monogamy.
12. Emily Washburn, "What to Know About Effective Altruism—Championed by Musk, Bankman-Fried, and Silicon Valley Giants," *Forbes*, March 8, 2023, https://www.forbes.com/sites/emilywashburn/2023/03/08/what-to-know-about-effective-altruism-championed-by-musk-bankman-fried-and-silicon-valley-giants/?sh=715772362d1e, accessed November 23, 2023; Brian Fung, "Jeff Bezos Says He Will Give Most of His Money to Charity," CNN, November 14, 2022, https://www.cnn.com/2022/11/14/business/jeff-bezos-charity/index.html; "About: The Giving Pledge," https://givingpledge.org/about, accessed November 23, 2023.
13. Dawkins, "Twelve Misunderstandings of Kin Selection"; Grafen, "Modelling in Behavioural Ecology."
14. Darwin, *The Descent of Man, and Selection in Relation to Sex*.
15. Pregnant Then Screwed, "A Third of New Parents Can Not Afford to Have More Children," November 15, 2022, https://pregnantthenscrewed.com/press-release-a-third-of-new-parents-can-not-afford-to-have-more-children/.
16. Jenae Barnes, "Elon Musk Isn't the Only Billionaire with 9-Plus Kids. Meet the U.S.' Richest People with the Most Children," *Forbes*, July 10, 2022,

https://www.forbes.com/sites/jenaebarnes/2022/07/10/elon-musk-isnt-the-only-billionaire-with-9-plus-kids-meet-the-us-richest-people-with-the-most-children/, accessed November 23, 2023; Julia Black, "Billionaires like Elon Musk Want to Save Civilization by Having Tons of Genetically Superior Kids. Inside the Movement to Take 'Control of Human Evolution,'" *Business Insider*, November 17, 2022, https://www.businessinsider.com/pronatalism-elon-musk-simone-malcolm-collins-underpopulation-breeding-tech-2022-11, accessed November 23, 2023.

17. Zerjal et al., "The Genetic Legacy of the Mongols."
18. Hammerstein and Noë, "Biological Trade and Markets."
19. Megan Twohey, "Kanye West and Adidas: How Misconduct Broke a Lucrative Partnership," *New York Times*, October 27, 2023, https://www.nytimes.com/2023/10/27/business/kanye-west-adidas-yeezy.html, accessed November 23, 2023.
20. Wiessner, "Risk, Reciprocity, and Social Influences on Kung San Economics."
21. Alvard and Nolin, "Rousseau's Whale Hunt?"
22. Micheletti et al., "Religious Celibacy Brings Inclusive Fitness Benefits."
23. Alexander, *The Biology of Moral Systems*. I am not making any claim about whether these interpretations are correct, but rather how the language of biology translates self-sacrifice. Motives are irrelevant in the biological sciences—but that doesn't mean they aren't important outside of biology.
24. Levon et al., "Accent Bias and Perceptions of Professional Competence in England."
25. For some people Francis might be better known for inventing the parseltongue language for the Harry Potter movies.
26. Janaya Wecker, "Celebrity Nepo Babies—Nepotism in Hollywood," *Cosmopolitan*, November 10, 2023, https://www.cosmopolitan.com/uk/entertainment/g45804095/nepo-babies-in-hollywood-list/, accessed November 23, 2023.
27. *The Mating Game* documentary (2021), shows how this manifests itself across species. Successful reproduction doesn't always involve the individual, though, given that inclusive fitness benefits can outweigh direct fitness in some cases.
28. Hamilton and Zuk, "Heritable True Fitness and Bright Birds."
29. Maynard-Smith and Harper, *Animal Signals*; Ghislandi et al., "Silk Wrapping of Nuptial Gifts Aids Cheating Behaviour in Male Spiders."
30. Rathore, Isvaran, and Guttal, "Lekking as Collective Behaviour."
31. Winegard, Winegard, and Geary, "The Status Competition Model of Cultural Production"; Beckwith, "Niger's Wodaabe."

32. Swami et al., "Factors Influencing Preferences for Height."

33. Interestingly, human reproductive strategies—how and when we rear children—mirror behaviors seen in local non-human animals worldwide.

34. Smith, Bird, and Bird, "The Benefits of Costly Signaling."

35. Bongaarts, "Human Population Growth and the Demographic Transition."

36. Kaplan et al., "A Theory of Human Life History Evolution."

37. Barsbai, Lukas, and Pondorfer, "Local Convergence of Behavior Across Species."

38. Bowles, Smith, and Borgerhoff Mulder, "The Emergence and Persistence of Inequality in Premodern Societies." Respectively, these categories refer to people who hunt and forage for food, who grow their food, who survive off cattle they herd, and who create larger farms for food production, usually relying on plows and animal labor. The distinctions aren't always clear in specific cases—for example, some societies have hunting and farming—and there is no assumption that one form of subsistence is better than any other.

39. Singh, "How Dowries Are Fuelling a Femicide Epidemic"; Akurugu, Dery, and Domanban, "Marriage, Bridewealth, and Power."

40. As of September 2023, Microsoft Word suggests I've made a spelling error with "matrilineality," and suggests "patrilineality" instead.

41. The journalist Gideon Rachman discusses the modern evolution of strongmen in his recent book, *The Age of the Strongman*.

42. Ronson, *The Psychopath Test*.

43. von Hippel and Trivers, "The Evolution and Psychology of Self-Deception"; Henrich, "Cultural Group Selection, Coevolutionary Processes, and Large-Scale Cooperation."

44. Goodman and Ewald, "The Evolution of Barriers to Exploitation."

45. Vollrath, "Uncoupling Elephant TP53 and Cancer"; Sulak, Fong, and Mika, "TP53 Copy Number Expansion Is Associated with the Evolution of Increased Body Size and an Enhanced DNA Damage Response in Elephants."

46. Miller et al., "Insect-Induced Conifer Defense"; Bouwmeester, "Dissecting the Pine Tree Green Chemical Factory."

47. Raffa, Powell, and Townsend, "Temperature-Driven Range Expansion of an Irruptive Insect Heightened by Weakly Coevolved Plant Defenses."

48. Dominey, "Female Mimicry in Male Bluegill Sunfish—A Genetic Polymorphism?"

49. Croston and Hauber, "The Ecology of Avian Brood Parasitism."

50. Schmid and Patrick, "The Trouble with Relying on How People Speak to Determine Asylum Cases."

51. Barker et al., "Cultural Transmission of Vocal Dialect in the Naked Mole-Rat."

52. The English word "cuckold" derives from cuckoo, and historically was applied to men who unknowingly raised the children of another person.

53. Strassmann, "The Function of Menstrual Taboos Among the Dogon."

54. While men directly benefit, evolutionarily speaking, from these practices, women can have an interest in perpetuating them, too: a potential grandmother, for example, will want to ensure her son's children are his own.

55. Walker, Flinn, and Hill, "Evolutionary History of Partible Paternity in Lowland South America."

56. Julius Caesar, *Gallic Wars*, Book 5, Chapter 14.

57. Starkweather and Hames, "A Survey of Non-Classical Polyandry."

58. Garfield et al., "Norm Violations and Punishments Across Human Societies."

59. There are a lot of definitions for the term "norm" in the literature, but a useful one is in the recent ethnographic comparison of forms of punishment: "suites of group-typical beliefs about what constitutes appropriate behavior in a given context." See Garfield et al., "Norm Violations and Punishments Across Human Societies."

60. Inhorn and Brown, "The Anthropology of Infectious Disease."

61. Singh, "Subjective Selection and the Evolution of Complex Culture."

5
The Power of Darkness

1. Howard Kurtz, "Stranger Than Fiction," *Washington Post*, May 13, 1998, https://www.washingtonpost.com/archive/politics/1998/05/13/stranger-than-fiction/d317a388-e47c-4e94-9a12-268494526f76/.

2. Howard Kurtz, "At New Republic, the Agony of Deceit," *Washington Post*, June 12, 1998, https://www.washingtonpost.com/archive/lifestyle/1998/06/12/at-new-republic-the-agony-of-deceit/0363ac50-8c48-40e3-86ba-5b83ed4541a8/.

3. I couldn't verify this quote online but Lane very helpfully confirmed by phone that he had used those exact words.

4. David Brooks, "People Are More Generous Than You May Think," *New York Times*, August 31, 2023, https://www.nytimes.com/2023/08/31/opinion/human-nature-good-bad-generous.html; Dwyer et al., "Are People Generous When the Financial Stakes Are High?"

5. CoDa: A Machine-Readable History of Cooperation Research to Search and Select Studies for On-Demand Meta-Analysis.

6. Yuan et al., "Did Cooperation Among Strangers Decline in the United States?"
7. Henrich, *The WEIRDest People in the World*.
8. Henrich et al., "In Search of Homo Economicus."
9. Wiessner, "Experimental Games and Games of Life among the Ju/'hoan Bushmen."
10. Yamagishi, "Exit from the Group as an Individualistic Solution to the Free Rider Problem in the United States and Japan"; Yamagishi, "The Provision of a Sanctioning System in the United States and Japan." Yamagishi's Japanese heritage would not, on Henrich's view, qualify him as WEIRD, but given his academic training in the United States, I am including this study as "WEIRD" research.
11. Dana, Cain, and Dawes, "What You Don't Know Won't Hurt Me."
12. At least, from the second player.
13. Haley and Fessler, "Nobody's Watching?"
14. A number of researchers have reproduced this study, with mixed findings. A meta-analysis (a study of studies) suggested that the presence of eyes may just increase the amount of money a person gives, rather than increase the odds of giving over not. See Nettle et al., "The Watching Eyes Effect in the Dictator Game."
15. Brown et al., "Moral Credentialing and the Rationalization of Misconduct."
16. Monin and Miller, "Moral Credentials and the Expression of Prejudice."
17. Khan and Dhar, "Licensing Effect in Consumer Choice."
18. Brown et al., "Moral Credentialing and the Rationalization of Misconduct."
19. Oxoby and Spraggon, "Mine and Yours."
20. Brosnan and de Waal, "Monkeys Reject Unequal Pay."
21. The philosopher John Rawls makes the seminal case for distinguishing these elements of a rule in his essay "Two Concepts of Rules" from 1955.
22. Kevin Rawlinson, "National Insurance Increase Is Right and Fair, Says Sajid Javid," *Guardian*, April 6, 2022, https://www.theguardian.com/politics/2022/apr/06/national-insurance-increase-is-right-and-fair-says-sajid-javid. "Equal" here can have, according to Aristotle, two senses: arithmetic and geometric. "Flat" equality, where everyone, regardless of their circumstances, is arithmetic (and a bit unreasonable), while geometric equality gives more weight to the merit of each individual (harder working, more family members, and so forth). Anthony Price very kindly pointed out this distinction to me.
23. Falk and Szech, "Morals and Markets."
24. It's also possible—although the authors don't discuss this—that the

Notes to Pages 120–126

participants believed the experimenters were ethically responsible for the mice, not them. Stanley Milgram's well-known shock therapy experiments in the mid-twentieth century took this approach explicitly.

25. Antony Barnett, "Price of Cocaine Paid with Blood," *Guardian*, February 13, 2005, https://www.theguardian.com/world/2005/feb/13/drugsand alcohol.colombia, accessed August 11, 2024.

26. *In re Holocaust Victim Assets Litigation*, 302 F. Supp. 2d 59 (E.D.N.Y. 2004), https://casetext.com/case/in-re-holocaust-victim-assets-litigation, accessed November 28, 2023. My thanks to Adam LeBor for sharing Korman's opinion with me.

27. "Why Is Switzerland So Rich?" March 23, 2022, https://studyingin switzerland.com/why-is-switzerland-so-rich/.

28. Swift, *Tale of a Tub*.

29. "Faith Schools—Why Not?" In 2019 I joined Humanists U.K. as a "young humanist," thinking the movement was about promoting the wellbeing of people worldwide, as it was in the sixteenth century—a slightly out-of-date view.

30. Mark Oppenheimer, "Will Misogyny Bring Down the Atheist Movement?" *BuzzFeed*, September 12, 2014, https://www.buzzfeed.com/markop penheimer/will-misogyny-bring-down-the-atheist-movement.

31. P. Z. Myers, "What Do You Do When Someone Pulls the Pin and Hands You a Grenade?" Free Thought Blogs, August 9, 2013, https://free thoughtblogs.com/pharyngula/2013/08/08/what-do-you-do-when-some one-pulls-the-pin-and-hands-you-a-grenade/.

32. Olivia Blair, "Richard Dawkins Dropped from Science Event for Tweeting Video Mocking Feminists and Islamists," *Independent*, February 2, 2016, https://www.independent.co.uk/news/people/richard-dawkins-vdeo-twitter-necss-event-feminism-a6841161.html.

33. Christopher Hitchens, "Why Women Aren't Funny," *Vanity Fair*, January 1, 2007, https://www.vanityfair.com/culture/2007/01/hitchens200701.

34. Sam Harris, "I'm Not the Sexist Pig You're Looking For," Making Sense podcast, https://www.samharris.org/blog/im-not-the-sexist-pig-youre-looking-for, accessed November 28, 2023.

35. Ashley Reese, "How the Wing's Empire Was Built on Trauma, Racism, and Neglect," *Jezebel*, June 12, 2020, https://jezebel.com/how-the-wings-empire-was-built-on-trauma-racism-and-n-1844000985.

36. Eloise Hendy, "The Girlbosses Who Girlbossed Too Close to the Sun: The Demise of 'Women's Utopia' The Wing Was Long Overdue," *Independent*, September 3, 2022, https://www.the-independent.com/life-style/women/the-wing-closed-feminism-b2158492.html.

37. David Armstrong and Ryan Gabrielson, "St. Jude Hoards Billions

While Many of Its Families Drain Their Savings," *ProPublica,* November 12, 2021, https://www.propublica.org/article/st-jude-hoards-billions-while-many-of-its-families-drain-their-savings.

38. "Nonprofit Compensation Packages of $1 Million or More," *Charity Watch,* https://www.charitywatch.org/nonprofit-compensation-packages-of-1-million-or-more, accessed November 28, 2023.

39. Zeke Faux, "'Don't You Remember Me?' The Crypto Hell on the Other Side of a Spam Text," *Bloomberg,* August 17, 2023, https://www.bloomberg.com/news/features/2023-08-17/my-crypto-hell-journey-started-with-a-wrong-number-scam-text.

40. The philosopher Daniel Dennett called this way of thinking "Popperian," referring to the twentieth-century philosopher of science Karl Popper.

41. Tae Kim, "Goldman Sachs Asks in Biotech Research Report: 'Is Curing Patients a Sustainable Business Model?'" CNBC, April 11, 2018, https://www.cnbc.com/2018/04/11/goldman-asks-is-curing-patients-a-sustainable-business-model.html.

6
Electrification

1. Adam LeBor, "Goering's List," *Sunday Times,* November 22, 1998, the Sunday Times Historical Archive. I am grateful to Adam LeBor, author of the first exposé of Albert Goering, published in 1998, who shared documents about him with me.

2. Rabinowitch and Meltzoff, "Synchronized Movement Experience Enhances Peer Cooperation in Preschool Children"; Tunçgenç and Cohen, "Interpersonal Movement Synchrony Facilitates Pro-Social Behavior in Children's Peer-Play."

3. Boyd and Richerson, *Culture and the Evolutionary Process.*

4. Horner and Whiten, "Causal Knowledge and Imitation/Emulation Switching in Chimpanzees (Pan Troglodytes) and Children (Homo Sapiens)."

5. Humphrey, "Follow My Leader."

6. As quoted in Elon, "Scenes from a Marriage."

7. Tullock, *Social Dilemma.*

8. Rana Foroohar, "Workers Could Be the Ones to Regulate AI," *Financial Times,* October 2, 2023, https://www.ft.com/content/edd17fbc-b0aa-4d96-b7ec-382394d7c4f3.

9. "Our Foreign and Domestic Position and Party Tasks: Speech Delivered to the Moscow Gubernia Conference of the R.C.P.(B.), November 21, 1920," https://www.marxists.org/archive/lenin/works/1920/nov/21.htm, accessed November 28, 2023.

10. Also published as "Poverty" in English.
11. Zoschencko, "Electrification." I apologize for bringing everything back to Russian literature.
12. Ostrom, "Collective Action and the Evolution of Social Norms"; Hardin, "The Tragedy of the Commons."
13. Ostrom, "Common-Pool Resources and Institutions."
14. Feldstein, "Government Internet Shutdowns Are Changing."
15. Ostrom and Gardner, "Coping with Asymmetries in the Commons."
16. Corrêa, Kahn, and Freitas, "Perverse Incentives in Fishery Management."
17. International Consortium of Investigative Journalists, "Paradise Papers."
18. Lightner, Pisor, and Hagen, "In Need-Based Sharing, Sharing Is More Important than Need."
19. Kalyeena Makortoff, "UK Taxpayers Left Footing Bill as Number of Fraudulent Covid Loans Soars," *Guardian*, September 14, 2023, https://www.theguardian.com/business/2023/sep/14/uk-taxpayers-left-footing-bill-as-number-of-fraudulent-covid-loans-soars.
20. Mazar, Amir, and Ariely, "The Dishonesty of Honest People."
21. Dwyer et al., "Unconditional Cash Transfers Reduce Homelessness."
22. Singer, *The Expanding Circle*.
23. Claire Wrathall, "Revolving Sofas, a 100-Inch TV, and 75 Damien Hirsts: Inside London's Most Expensive Hotel Suite," *Financial Times*, October 19, 2023, https://www.ft.com/content/8ef079e4-e143-4b9b-91b1-de2a58748c4c.
24. Bucher, "Global Inequality Is a Failure of Imagination. Here's Why."
25. Wells, *The Metabolic Ghetto*.
26. Hochberg, "Developmental Plasticity in Child Growth and Maturation"; Singhal, "Long-Term Adverse Effects of Early Growth Acceleration or Catch-Up Growth."
27. Brunner, "Social Factors and Cardiovascular Morbidity." You can also view life expectancy as predicted by London Tube map stations at http://life.mappinglondon.co.uk/life/.
28. "Accounting for the Widening Mortality Gap Between American Adults with and without a BA."
29. Wells, *The Metabolic Ghetto*.
30. Wilkinson, "Income Inequality, Social Cohesion, and Health."
31. Francis-Devine, Danechi, and Malik, "Food Poverty."
32. Wells, "An Evolutionary Perspective on Social Inequality and Health Disparities."
33. Pierotti, "Gull-Puffin Interactions on Great Island, Newfoundland."

34. Emerson, *The Conduct of Life*.

35. "Violence: Our Deadly Epidemic and Its Causes | Office of Justice Programs."

36. John Burn-Murdoch, "Millennials Are Shattering the Oldest Rule in Politics," *Financial Times*, December 30, 2022, https://www.ft.com/content/c361e372-769e-45cd-a063-f5c0a7767cf4.

37. Nettle et al., "What Do British People Want from a Welfare System?"

38. Bowles, Smith, and Borgerhoff Mulder, "The Emergence and Persistence of Inequality in Premodern Societies."

39. "OECD Web Archive."

40. Katherine Schaeffer, "6 Facts About Economic Inequality in the U.S.," *Pew Research Center* (blog), https://www.pewresearch.org/short-reads/2020/02/07/6-facts-about-economic-inequality-in-the-u-s/, accessed November 29, 2023.

41. "Global Tax Evasion Report 2024."

42. Wells, *The Metabolic Ghetto*, 213–32.

43. Daniel Nettle, "What Do People Want from a Welfare System?" *Daniel Nettle* (blog), September 21, 2023, https://www.danielnettle.org.uk/2023/09/21/what-do-people-want-from-a-welfare-system/.

44. Goldenberg et al., "Vaccine Mandates and Public Trust Do Not Have to Be Antagonistic."

45. Ginges et al., "Sacred Bounds on Rational Resolution of Violent Political Conflict"; Atran, "The Devoted Actor."

46. Ginges et al., "Sacred Bounds on Rational Resolution of Violent Political Conflict."

47. "The Gold Standard," https://www.goldstandard.org/, accessed November 29, 2023.

48. Patrick Greenfield, "More Than 90% of Rainforest Carbon Offsets by Biggest Certifier Are Worthless, Analysis Shows," *Guardian*, January 18, 2023, https://www.theguardian.com/environment/2023/jan/18/revealed-forest-carbon-offsets-biggest-provider-worthless-verra-aoe.

49. This figure is taken from a press release: KPMG UK, "Over Half of UK Consumers Prepared to Boycott Brands over Misleading Green Claims," September 18, 2023, https://kpmg.com/uk/en/home/media/press-releases/2023/09/over-half-of-uk-consumers-prepared-to-boycott-brands-over-misleading-green-claims.html. I could not locate the original report, though, and e-mailed a representative at KPMG to verify the claim. Claire Barratt of KPMG UK kindly responded and sent over some relevant data. One question YouGov asked 2,067 people in the UK was as follows: "Please imagine a company has been found to have been misleading in their sustainability / green claims, or accused of greenwashing. . . . Assuming you were already

doing these activities, which, if any, of the following activities would you stop doing for the company?" Interestingly, while 54 percent of people said they would stop buying products/services from the company, only 21 percent said they would stop working at it—and only 38 percent said they would stop investing in it.

50. Ronson, *The Psychopath Test*.
51. Tullock, *Social Dilemma*.

7
We

1. Wittgenstein, *Philosophical Investigations*.
2. Bond and Depaulo, "Individual Differences in Judging Deception."
3. Bond and Depaulo, "Individual Differences in Judging Deception."
4. My thanks to Jo Wolff for bringing up the importance of making this distinction clear.
5. Derek Thompson, "The Republican War on College," *Atlantic*, November 20, 2017, https://www.theatlantic.com/business/archive/2017/11/republican-college/546308/.
6. Livingston et al., "Dementia Prevention, Intervention, and Care."
7. Roozenbeek et al., "Psychological Inoculation Improves Resilience Against Misinformation on Social Media."
8. David Spiegelhalter's *The Art of Statistics* is a particular recommendation.
9. Supran and Oreskes, "Assessing ExxonMobil's Climate Change Communications (1977–2014)."
10. See Trivers, *Deceit and Self-Deception*.
11. Williams, *Essays and Reviews*.
12. See O'Neill, "Linking Trust to Trustworthiness."
13. Gambetta, *Codes of the Underworld*. See also Hamill and Gambetta, *Streetwise*.
14. Tett, *Anthro-Vision*.
15. Beamer, Tau, and Vitousek, *Islands and Cultures*.
16. Sogbanmu et al., "Indigenous Youth Must Be at the Forefront of Climate Diplomacy."
17. "Igliniit Viewer," https://www.geos.ed.ac.uk/~mscgis/09-10/s0972776/background.html, accessed November 29, 2023.
18. Pizzol et al., "Relationship Between Severe Mental Illness and Physical Multimorbidity."
19. Sen, *The Idea of Justice*.
20. Friedlingstein et al., "Global Carbon Budget 2022."

21. Oliver Milman, "Governments Falling Woefully Short of Paris Climate Pledges, Study Finds," *Guardian*, September 15, 2021, https://www.theguardian.com/science/2021/sep/15/governments-falling-short-paris-climate-pledges-study.

22. To me this accords coincidentally with Aristotle's view in his *Nicomachean Ethics:* "For the beginning (starting point) is 'the that,' and if this is sufficiently apparent to a person he will not in addition have a need for 'the because.' Such a person has, or can easily get hold of, beginnings (starting points), whereas he who has neither, let him harken to the words of Hesiod: 'The best man of all is he who knows everything himself, Good also the man who accepts another's sound advice; But the man who neither knows himself nor takes to heart What another says, he is no good at all.'" (I.4, 1095b1-13), 34. "The that" encompasses norms or behavioral rules; "the because" is the moral foundations of those roots.

23. Mirrlees, *Lud-in-the-Mist*.

Bibliography

Acemoglu, Daron, and James A. Robinson. *Why Nations Fail*. London: Profile, 2013.

Aktipis, C. Athena. *The Cheating Cell*. Princeton: Princeton University Press, 2020. https://press.princeton.edu/books/hardcover/9780691163840/the-cheating-cell.

Aktipis, C. A., L. Cronk, and R. Aguiar. "Risk-Pooling and Herd Survival: An Agent-Based Model of a Maasai Gift-Giving System." *Human Ecology* 39, no. 2 (2011): 131–40. https://doi.org/10.1007/s10745-010-9364-9.

Akurugu, Constance Awinpoka, Isaac Dery, and Paul Bata Domanban. "Marriage, Bridewealth, and Power: Critical Reflections on Women's Autonomy Across Settings in Africa." *Evolutionary Human Sciences* 4 (January 2022): e30. https://doi.org/10.1017/ehs.2022.27.

Alexander, Richard D. *The Biology of Moral Systems*. Hawthorne, N.Y.: Aldine de Gruyter, 1987.

Alvard, Michael S., and David A. Nolin. "Rousseau's Whale Hunt? Coordination Among Big-Game Hunters." *Current Anthropology* 43, no. 4 (2002): 533–59. https://doi.org/10.1086/341653.

Alvarez, R. Michael, Ramit Debnath, and Daniel Ebanks. "Why Don't Americans Trust University Researchers and Why It Matters for Climate Change." *PLOS Climate* 2, no. 9 (September 6, 2023): e0000147. https://doi.org/10.1371/journal.pclm.0000147.

Armstrong, W. A. *Rossel Island: An Ethnological Study*. Cambridge: Cambridge University Press, 2011; originally published in 1928.

Atran, Scott. "The Devoted Actor: Unconditional Commitment and Intractable Conflict Across Cultures." *Current Anthropology* 57, no. S13 (June 2016): S192–203. https://doi.org/10.1086/685495.

Axelrod, Robert. *The Evolution of Cooperation.* Revised edition. New York: Basic, 2006.

Axelrod, Robert, and William D. Hamilton. "The Evolution of Cooperation." *Science* 211, no. 4489 (1981): 1390–96. https://doi.org/10.1126/science.7466396.

Barker, Alison J., Grigorii Veviurko, Nigel C. Bennett, Daniel W. Hart, Lina Mograby, and Gary R. Lewin. "Cultural Transmission of Vocal Dialect in the Naked Mole-Rat." *Science* 371, no. 6528 (January 29, 2021): 503–7. https://doi.org/10.1126/science.abc6588.

Barsbai, Toman, Dieter Lukas, and Andreas Pondorfer. "Local Convergence of Behavior Across Species." *Science* 371, no. 6526 (January 15, 2021): 292–95. https://doi.org/10.1126/science.abb7481.

Beamer, Kamanamaikalani, Te Maire Tau, and Peter M. Vitousek. *Islands and Cultures: How Pacific Islands Provide Paths Toward Sustainability.* New Haven: Yale University Press, 2023.

Beckwith, C. "Niger's Wodaabe: People of the Taboo." *National Geographic* 164, no. 4 (1983): 483–509.

Berger, Lee R., Tebogo Makhubela, Keneiloe Molopyane, Ashley Krüger, Patrick Randolph-Quinney, Marina Elliott, Becca Peixotto, et al. "Evidence for Deliberate Burial of the Dead by Homo Naledi." *eLife* 12 (July 12, 2023). https://doi.org/10.7554/eLife.89106.

Bond, C. F., and B. M. Depaulo. "Individual Differences in Judging Deception: Accuracy and Bias." *Psychological Bulletin* 134, no. 4 (2008): 477–92. https://doi.org/10.1037/0033-2909.134.4.477.

Bongaarts, John. "Human Population Growth and the Demographic Transition." *Philosophical Transactions of the Royal Society B: Biological Sciences* 364, no. 1532 (October 27, 2009): 2985–90. https://doi.org/10.1098/rstb.2009.0137.

Bouwmeester, Harro. "Dissecting the Pine Tree Green Chemical Factory." *Journal of Experimental Botany* 70, no. 1 (January 1, 2019): 4–6. https://doi.org/10.1093/jxb/ery407.

Bowles, Samuel, and Herbert Gintis. "Homo Reciprocans." *Nature* 415, no. 6868 (January 2002): 125–27. https://doi.org/10.1038/415125a.

Bowles, Samuel, Eric Alden Smith, and Monique Borgerhoff Mulder. "The Emergence and Persistence of Inequality in Premodern Societies: Introduction to the Special Section." *Current Anthropology* 51, no. 1 (February 2010): 7–17. https://doi.org/10.1086/649206.

Boyd, Robert, and Peter J. Richerson. *Culture and the Evolutionary Process.* 2nd edition. Chicago: University of Chicago Press, 1988.

Brosnan, Sarah F., and Frans B. M. de Waal. "Monkeys Reject Unequal Pay."

Nature 425, no. 6955 (September 2003): 297–99. https://doi.org/10.1038/nature01963.

Brown, Ryan P., Michael Tamborski, Xiaoqian Wang, Collin D. Barnes, Michael D. Mumford, Shane Connelly, and Lynn D. Devenport. "Moral Credentialing and the Rationalization of Misconduct." *Ethics & Behavior* 21, no. 1 (January 2011): 1–12. https://doi.org/10.1080/10508422.2011.537566.

Brunner, Eric John. "Social Factors and Cardiovascular Morbidity." *Neuroscience and Biobehavioral Reviews* 74, no. Pt B (March 2017): 260–68. https://doi.org/10.1016/j.neubiorev.2016.05.004.

Bucher, Gabriela. "Global Inequality Is a Failure of Imagination. Here's Why." *World Economic Forum*. January 16, 2023. https://www.weforum.org/agenda/2023/01/global-inequality-is-a-failure-of-imagination/.

Burnyeat, M. F., ed. "Aristotle on Learning to Be Good." In *Explorations in Ancient and Modern Philosophy*, 2:259–81. Cambridge: Cambridge University Press, 2012. https://doi.org/10.1017/CBO9780511974069.016.

Byrne, R. W., and A. Whiten. "Cognitive Evolution in Primates: Evidence from Tactical Deception." *Man* 27, no. 3 (1992): 609–27. https://doi.org/10.2307/2803931.

"C. Julius Caesar, Gallic War, Book 5, Chapter 14." https://www.perseus.tufts.edu/hopper/text?doc=Perseus%3Atext%3A1999.02.0001%3Abook%3D5%3Achapter%3D14, accessed November 26, 2023.

Case, Anne, and Angus Deaton. "Accounting for the Widening Mortality Gap Between American Adults With and Without a BA." *Brookings Papers on Economic Activity*, Fall 2023. https://www.brookings.edu/articles/accounting-for-the-widening-mortality-gap-between-american-adults-with-and-without-a-ba/, accessed November 29, 2023.

Chagnon, Napoleon A., and William Irons. *Evolutionary Biology and Human Social Behavior: An Anthropological Perspective*. North Scituate, Mass.: Duxbury, 1979.

Chartrand, T. L., and J. A. Bargh. "The Chameleon Effect: The Perception-Behavior Link and Social Interaction." *Journal of Personality and Social Psychology* 76, no. 6 (June 1999): 893–910. https://doi.org/10.1037//0022-3514.76.6.893.

Chaudhary, Nikhil, Gul Deniz Salali, James Thompson, Aude Rey, Pascale Gerbault, Edward Geoffrey Jedediah Stevenson, Mark Dyble, et al. "Competition for Cooperation: Variability, Benefits and Heritability of Relational Wealth in Hunter-Gatherers." *Scientific Reports* 6, no. 1 (July 12, 2016): 29120. https://doi.org/10.1038/srep29120.

Clark, Cory J., Lee Jussim, Komi Frey, Sean T. Stevens, Musa al-Gharbi, Karl Aquino, J. Michael Bailey, et al. "Prosocial Motives Underlie Scientific

Censorship by Scientists: A Perspective and Research Agenda." *Proceedings of the National Academy of Sciences* 120, no. 48 (November 28, 2023): e2301642120. https://doi.org/10.1073/pnas.2301642120.

CoDa: A Machine-Readable History of Cooperation Research to Search and Select Studies for On-Demand Meta-Analysis. Spadaro, G., I. Tiddi, S. Columbus, S. Jin, A. Ten Teije, CoDa Team, & D. Balliet. "The Cooperation Databank: Machine-Readable Science Accelerates Research Synthesis." *Perspectives on Psychological Science* 17, no. 5 (2022): 1472–89. https://cooperationdatabank.org/, accessed November 27, 2023.

Corrêa, Maria Angélica de Almeida, James R. Kahn, and Carlos Edwar de Carvalho Freitas. "Perverse Incentives in Fishery Management: The Case of the Defeso in the Brazilian Amazon." *Ecological Economics* 106 (October 1, 2014): 186–94. https://doi.org/10.1016/j.ecolecon.2014.07.023.

Croston, Rebecca, and Mark E. Hauber. "The Ecology of Avian Brood Parasitism." *Nature Education Knowledge* 3, no. 10 (2010): 56. https://www.nature.com/scitable/knowledge/library/the-ecology-of-avian-brood-parasitism-14724491/.

Dana, Jason, Daylian M. Cain, and Robyn M. Dawes. "What You Don't Know Won't Hurt Me: Costly (but Quiet) Exit in Dictator Games." *Organizational Behavior and Human Decision Processes* 100, no. 2 (July 1, 2006): 193–201. https://doi.org/10.1016/j.obhdp.2005.10.001.

Darwin, Charles R. *The Descent of Man, and Selection in Relation to Sex*, 1871. https://doi.org/10.1038/011305a0.

Dawkins, Richard. *The Blind Watchmaker: Why the Evidence of Evolution Reveals a Universe Without Design*. Illustrated edition. New York: W. W. Norton, 2018.

Dawkins, Richard. "Twelve Misunderstandings of Kin Selection." *Zeitschrift für Tierpsychologie* 51, no. 2 (1979): 184–200. https://doi.org/10.1111/j.1439-0310.1979.tb00682.x.

DeMerchant, Kendra, and Sean P. Roach. "Vocal Responses of Hermit Thrush (Catharus Guttatus) Males to Territorial Playback of Conspecific Song." *Ibis* 164, no. 3 (2022): 793–99. https://doi.org/10.1111/ibi.13044.

Dey, Cody J., James Dale, and James S. Quinn. "Manipulating the Appearance of a Badge of Status Causes Changes in True Badge Expression." *Proceedings of the Royal Society B: Biological Sciences* 281, no. 1775 (January 22, 2014): 20132680. https://doi.org/10.1098/rspb.2013.2680.

Dominey, W. "Female Mimicry in Male Bluegill Sunfish: A Genetic Polymorphism?" *Nature* 284 (1980): 546–48. https://doi.org/10.1038/284546a0.

Dunbar, R. I. M. "Coevolution of Neocortical Size, Group Size, and Language in Humans." *Behavioral and Brain Sciences* 16, no. 4 (December 1993): 681–94. https://doi.org/10.1017/S0140525X00032325.

Dunbar, Robin. *Grooming, Gossip, and the Evolution of Language.* Cambridge: Harvard University Press, 1998. https://www.hup.harvard.edu/catalog.php?isbn=9780674363366.

Dwyer, Ryan J., William J. Brady, Chris Anderson, and Elizabeth W. Dunn. "Are People Generous When the Financial Stakes Are High?" *Psychological Science* 34, no. 9 (September 1, 2023): 999–1006. https://doi.org/10.1177/09567976231184887.

Dwyer, Ryan, Anita Palepu, Claire Williams, Daniel Daly-Grafstein, and Jiaying Zhao. "Unconditional Cash Transfers Reduce Homelessness." *Proceedings of the National Academy of Sciences* 120, no. 36 (September 5, 2023): e2222103120. https://doi.org/10.1073/pnas.2222103120.

Elon, Amos. "Scenes from a Marriage." *New York Review of Books,* July 5, 2001. https://www.nybooks.com/articles/2001/07/05/scenes-from-a-marriage/.

Emerson, Ralph Waldo. *The Conduct of Life.* Edited by Alba Longa. CreateSpace Independent Publishing Platform, 2016.

Erasmus. *In Praise of Folly.* London: Reeves & Turner, 1876. https://www.gutenberg.org/files/30201/30201-h/30201-h.htm, accessed November 23, 2023.

Eutax. "Global Tax Evasion Report 2024." https://www.taxobservatory.eu/publication/global-tax-evasion-report-2024/, accessed November 29, 2023.

Falk, Armin, and Nora Szech. "Morals and Markets." *Science* 340, no. 6133 (May 10, 2013): 707–11. https://doi.org/10.1126/science.1231566.

Feldstein, Steven. "Government Internet Shutdowns Are Changing. How Should Citizens and Democracies Respond?" Carnegie Endowment for International Peace. https://carnegieendowment.org/2022/03/31/government-internet-shutdowns-are-changing.-how-should-citizens-and-democracies-respond-pub-86687, accessed November 28, 2023.

Fisher, Irving. "What Is Capital?" *Economic Journal* 6, no. 24 (1896): 509–34. https://doi.org/10.2307/2957184.

Fonseca, Miguel A., and Kim Peters. "Is It Costly to Deceive? People Are Adept at Detecting Gossipers' Lies but May Not Reward Honesty." *Philosophical Transactions of the Royal Society of London. Series B, Biological Sciences* 376, no. 1838 (November 22, 2021): 20200304. https://doi.org/10.1098/rstb.2020.0304.

Francis-Devine, Brigid, Shadi Danechi, and Xameerah Malik. "Food Poverty: Households, Food Banks, and Free School Meals." House of Commons Library, U.K. Parliament. November 28, 2023. https://commonslibrary.parliament.uk/research-briefings/cbp-9209/.

Friedlingstein, Pierre, Michael O'Sullivan, Matthew W. Jones, Robbie M.

Andrew, Luke Gregor, Judith Hauck, Corinne Le Quéré, et al. "Global Carbon Budget 2022." *Earth System Science Data* 14, no. 11 (November 11, 2022): 4811–4900. https://doi.org/10.5194/essd-14-4811-2022.

Full Fact. "ONS Data Shows Lower Death Rates in People Vaccinated Against Covid-19," 14:15:07.165531+00:00. https://fullfact.org/health/covid-vaccines-ONS-florida/.

Garfield, Zachary H., Erik J. Ringen, William Buckner, Dithapelo Medupe, Richard W. Wrangham, and Luke Glowacki. "Norm Violations and Punishments Across Human Societies." *Evolutionary Human Sciences* 5 (January 2023): e11. https://doi.org/10.1017/ehs.2023.7.

Garfield, Zachary H., Ryan Schacht, Emily R. Post, Dominique Ingram, Andrea Uehling, and Shane J. Macfarlan. "The Content and Structure of Reputation Domains Across Human Societies: A View from the Evolutionary Social Sciences." *Philosophical Transactions of the Royal Society B: Biological Sciences* 376, no. 1838 (November 22, 2021): 20200296. https://doi.org/10.1098/rstb.2020.0296.

Ghislandi, Paolo Giovanni, Michelle Beyer, Patricia Velado, and Cristina Tuni. "Silk Wrapping of Nuptial Gifts Aids Cheating Behaviour in Male Spiders." *Behavioral Ecology* 28, no. 3 (May 1, 2017): 744–49. https://doi.org/10.1093/beheco/arx028.

Ginges, Jeremy, Scott Atran, Douglas Medin, and Khalil Shikaki. "Sacred Bounds on Rational Resolution of Violent Political Conflict." *Proceedings of the National Academy of Sciences* 104, no. 18 (May 2007): 7357–60. https://doi.org/10.1073/pnas.0701768104.

Goldenberg, Maya J., Bipin Adhikari, Lorenz von Seidlein, Phaik Yeong Cheah, and Heidi J. Larson. "Vaccine Mandates and Public Trust Do Not Have to Be Antagonistic." *Nature Human Behaviour* 7, no. 10 (October 2023): 1605–6. https://doi.org/10.1038/s41562-023-01720-8.

Goodman, Jonathan R. "The Problem of Opportunity." *Biology & Philosophy* 38, no. 6 (November 7, 2023): 48. https://doi.org/10.1007/s10539-023-09936-8.

Goodman, Jonathan R., and Paul W. Ewald. "The Evolution of Barriers to Exploitation: Sometimes the Red Queen Can Take a Break." *Evolutionary Applications* 14, no. 9 (September 2021): 2179–88. https://doi.org/10.1111/eva.13280.

Grafen, Alan. "Biological Signals as Handicaps." *Journal of Theoretical Biology* 144, no. 4 (June 21, 1990): 517–46. https://doi.org/10.1016/S0022-5193(05)80088-8.

Haley, Kevin J., and Daniel M. T. Fessler. "Nobody's Watching? Subtle Cues Affect Generosity in an Anonymous Economic Game." *Evolution and*

Bibliography 207

Human Behavior 26, no. 3 (2005): 245–56. https://doi.org/10.1016/j.evol humbehav.2005.01.002.

Hamilton, W. D. "The Genetical Evolution of Social Behaviour." I. *Journal of Theoretical Biology* 7, no. 1 (1964): 1–16. https://doi.org/10.1016/0022-5193 (64)90038-4.

Hamilton, W. D., and M. Zuk. "Heritable True Fitness and Bright Birds: A Role for Parasites?" *Science* 218, no. 4570 (October 22, 1982): 384–87. https://doi.org/10.1126/science.7123238.

Hammerstein, P., and R. Noë. "Biological Trade and Markets." *Philosophical Transactions of the Royal Society of London. Series B, Biological Sciences* 371, no. 1687 (2016): 20150101. https://doi.org/10.1098/rstb.2015.0101.

Hardin, Garrett. "The Tragedy of the Commons." *Science* 162, no. 3859 (1968): 1243–48.

Hare, Brian. "Survival of the Friendliest: Homo Sapiens Evolved via Selection for Prosociality." *Annual Review of Psychology* 68 (January 3, 2017): 155–86. https://doi.org/10.1146/annurev-psych-010416-044201.

Haridy, Yara, Florian Witzmann, Patrick Asbach, Rainer R. Schoch, Nadia Fröbisch, and Bruce M. Rothschild. "Triassic Cancer: Osteosarcoma in a 240-Million-Year-Old Stem-Turtle." *JAMA Oncology* 5, no. 3 (March 1, 2019): 425–26. https://doi.org/10.1001/jamaoncol.2018.6766.

Harmand, Sonia, Jason E. Lewis, Craig S. Feibel, Christopher J. Lepre, Sandrine Prat, Arnaud Lenoble, Xavier Boës, et al. "3.3-Million-Year-Old Stone Tools from Lomekwi 3, West Turkana, Kenya." *Nature* 521, no. 7552 (May 2015): 310–15. https://doi.org/10.1038/nature14464.

Henrich, Joseph. "Cultural Group Selection, Coevolutionary Processes and Large-Scale Cooperation." *Journal of Economic Behavior and Organization* 53, no. 1 (January 1, 2004): 3–35, Evolution and Altruism special issue. https://doi.org/10.1016/S0167-2681(03)00094-5.

Henrich, Joseph. *The WEIRDest People in the World: How the West Became Psychologically Peculiar and Particularly Prosperous*. New York: Farrar, Straus and Giroux, 2020. https://us.macmillan.com/books/9780374710453 /theweirdestpeopleintheworld.

Henrich, Joseph, Robert Boyd, Samuel Bowles, Colin Camerer, Ernst Fehr, Herbert Gintis, and Richard McElreath. "In Search of Homo Economicus: Behavioral Experiments in 15 Small-Scale Societies." *American Economic Review* 91, no. 2 (May 2001): 73–78. https://doi.org/10.1257/aer .91.2.73.

Hill, J. "Prestige and Reproductive Success in Man." *Ethology and Sociobiology* 5, no. 2 (January 1, 1984): 77–95. https://doi.org/10.1016/0162-3095(84) 90011-6.

Hill, Kim, and A. Magdalena Hurtado. "Cooperative Breeding in South American Hunter-Gatherers." *Proceedings of the Royal Society B: Biological Sciences* 276, no. 1674 (August 19, 2009): 3863–70. https://doi.org/10.1098/rspb.2009.1061.

Hippel, William von, and Robert Trivers. "The Evolution and Psychology of Self-Deception." *The Behavioral and Brain Sciences* 34, no. 1 (February 2011): 1–16; discussion 16–56. https://doi.org/10.1017/S0140525X10001354.

Hochberg, Ze'ev. "Developmental Plasticity in Child Growth and Maturation." *Frontiers in Endocrinology* 2 (September 29, 2011): 41. https://doi.org/10.3389/fendo.2011.00041.

Holmberg, Allan R. *Nomads of the Long Bow: The Siriono of Eastern Bolivia.* 1969; reprint, London: Forgotten Books, 2018.

Horner, Victoria, and Andrew Whiten. "Causal Knowledge and Imitation/Emulation Switching in Chimpanzees (Pan Troglodytes) and Children (Homo Sapiens)." *Animal Cognition* 8, no. 3 (July 2005): 164–81. https://doi.org/10.1007/s10071-004-0239-6.

Humanists UK. "Faith Schools—Why Not?" https://humanists.uk/campaigns/schools-and-education/faith-schools/faith-schools-why-not/, accessed November 28, 2023.

"Hume Texts Online." https://davidhume.org/texts/m/d, accessed November 23, 2023.

Humphrey, N. K. "The Social Function of Intellect." In *Growing Points in Ethology.* Cambridge: Cambridge University Press, 1976.

Humphrey, Nicholas. "Follow My Leader." In *The Mind Made Flesh: Essays from the Frontiers of Evolution and Psychology,* edited by Nicholas Humphrey, 330–39. Oxford University Press, 2002. https://web-archive.southampton.ac.uk/cogprints.org/2218/index.html.

Inhorn, Marcia C., and Peter J. Brown. "The Anthropology of Infectious Disease." *Annual Review of Anthropology* 19 (1990): 89–117.

International Consortium of Investigative Journalists. "Paradise Papers: Secrets of the Global Elite," November 5, 2017. https://www.icij.org/investigations/paradise-papers/.

"The Internet Classics Archive | The Republic by Plato." http://classics.mit.edu/Plato/republic.3.ii.html, accessed November 22, 2023.

James, Gary D. "Bergmann's and Allen's Rules." In *The International Encyclopedia of Biological Anthropology,* 1–3. John Wiley & Sons, 2018. https://doi.org/10.1002/9781118584538.ieba0048.

Kaplan, Hillard, Kim Hill, Jane Lancaster, and A. Magdalena Hurtado. "A Theory of Human Life History Evolution: Diet, Intelligence, and Longevity." *Evolutionary Anthropology: Issues, News, and Reviews* 9, no. 4 (2000): 156–85. https://doi.org/10.1002/1520-6505(2000)9:4<156::AID-EVAN5>3.0.CO;2-7.

Bibliography

Khan, Uzma, and Ravi Dhar. "Licensing Effect in Consumer Choice." *Journal of Marketing Research* 43, no. 2 (2006), American Marketing Association. https://journals.sagepub.com/doi/full/10.1509/jmkr.43.2.259?casa_token=Mk6qHMtFyp8AAAAA%3AmhnnJx1brdy_jxcnGZLy_mxvYrKPncLooZU8q_M4lHt6lGwHyZ_jgSwK6xt4Zdnokn8W_Q5Kbo8, accessed November 27, 2023.

Koenig, Walter D., Philip N. Hooge, Mark T. Stanback, and Joseph Haydock. "Natal Dispersal in the Cooperatively Breeding Acorn Woodpecker." *The Condor* 102, no. 3 (2000): 492–502. https://doi.org/10.2307/1369780.

Kropotkin, Kniaz Petr Alekseevich. *Mutual Aid: A Factor of Evolution*. 1902. American Psychological Association, APA PsycNet. https://psycnet.apa.org/record/1902-10282-000.

Lackner, Simone, Frederico Francisco, Cristina Mendonça, André Mata, and Joana Gonçalves-Sá. "Intermediate Levels of Scientific Knowledge Are Associated with Overconfidence and Negative Attitudes Towards Science." *Nature Human Behaviour* 7, no. 9 (September 2023): 1490–1501. https://doi.org/10.1038/s41562-023-01677-8.

Lee, Richard B., and Irven DeVore, eds. *Kalahari Hunter-Gatherers: Studies of the !Kung San and Their Neighbors*. Cambridge: Harvard University Press, 1976.

Levine, Nancy E., and Joan B. Silk. "Why Polyandry Fails: Sources of Instability in Polyandrous Marriages." *Current Anthropology* 38, no. 3 (June 1997): 375–98. https://doi.org/10.1086/204624.

Levinson, Stephen C. "Introduction: The Evolution of Culture in a Microcosm," October 7, 2005. https://doi.org/10.7551/mitpress/2870.003.0004.

Levon, Erez, Devyani Sharma, Dominic J. L. Watt, Amanda Cardoso, and Yang Ye. "Accent Bias and Perceptions of Professional Competence in England." *Journal of English Linguistics* 49, no. 4 (December 1, 2021): 355–88. https://doi.org/10.1177/00754242211046316.

Lightner, Aaron D., Anne C. Pisor, and Edward H. Hagen. "In Need-Based Sharing, Sharing Is More Important than Need." *Evolution and Human Behavior*, Special Issue: Dispatches from the Field Part I, 44, no. 5 (September 1, 2023): 474–84. https://doi.org/10.1016/j.evolhumbehav.2023.02.010.

Livingston, Gill, Jonathan Huntley, Andrew Sommerlad, David Ames, Clive Ballard, Sube Banerjee, Carol Brayne, et al. "Dementia Prevention, Intervention, and Care: 2020 Report of the Lancet Commission." *The Lancet* 396, no. 10248 (August 8, 2020): 413–46. https://doi.org/10.1016/S0140-6736(20)30367-6.

Lonsdorf, Elizabeth V. "Sex Differences in the Development of Termite-Fishing Skills in the Wild Chimpanzees, Pan Troglodytes Schweinfurthii, of Gombe National Park, Tanzania." *Animal Behaviour* 70, no. 3 (September 1, 2005): 673–83. https://doi.org/10.1016/j.anbehav.2004.12.014.

Mauss, M. *The Gift: The Form and Reason for Exchange in Archaic Societies.* London and New York: Routledge Classics, 1925.

Maynard-Smith, John, and David Harper. *Animal Signals.* Oxford Series in Ecology and Evolution. Oxford: Oxford University Press, 2003.

Mazar, Nina, On Amir, and Dan Ariely. "The Dishonesty of Honest People: A Theory of Self-Concept Maintenance." *Journal of Marketing Research* 45, no. 6 (December 1, 2008): 633–44. https://doi.org/10.1509/jmkr.45.6.633.

Micheletti, Alberto J. C., Erhao Ge, Liqiong Zhou, Yuan Chen, Hanzhi Zhang, Juan Du, and Ruth Mace. "Religious Celibacy Brings Inclusive Fitness Benefits." *Proceedings of the Royal Society B: Biological Sciences* 289, no. 1977 (June 22, 2022): 20220965. https://doi.org/10.1098/rspb.2022.0965.

Milgram, Stanley, and Philip Zimbardo. *Obedience to Authority: An Experimental View.* London: Pinter & Martin, 2010.

Miller, Barbara, Lufiani L. Madilao, Steven Ralph, and Jörg Bohlmann. "Insect-Induced Conifer Defense: White Pine Weevil and Methyl Jasmonate Induce Traumatic Resinosis, de Novo Formed Volatile Emissions, and Accumulation of Terpenoid Synthase and Putative Octadecanoid Pathway Transcripts in Sitka Spruce." *Plant Physiology* 137, no. 1 (January 2005): 369–82. https://doi.org/10.1104/pp.104.050187.

Mills, Cynthia. "Unusual Suspects." *The Sciences* 37, no. 4 (July 1, 1997): 32–37.

Mirrlees, Hope. *Lud-in-the-Mist.* London: Millennium, 2000.

Monin, B., and D. T. Miller. "Moral Credentials and the Expression of Prejudice." *Journal of Personality and Social Psychology* 81, no. 1 (July 2001): 33–43.

Montaigne, Michel de. *Essays,* translated by John M. Cohen, revised edition. Penguin, 1993.

Moore, John H. "The Evolution of Exploitation." *Critique of Anthropology* 2, no. 8 (February 1, 1977): 33–48. https://doi.org/10.1177/0308275X7700200803.

Moore, John H. "The Exploitation of Women in Evolutionary Perspective." *Critique of Anthropology* 3, no. 9–10 (January 1, 1978): 83–100. https://doi.org/10.1177/0308275X7800300904.

Nesse, Randolph M. "Social Selection Is a Powerful Explanation for Prosociality." *Behavioral and Brain Sciences* 39 (2016): e47. https://doi.org/10.1017/S0140525X15000308.

Nettle, Daniel, Joe Chrisp, Elliott Johnson, and Matthew T. Johnson. "What

Bibliography

Do British People Want from a Welfare System? Conjoint Survey Evidence on Generosity, Conditionality, Funding, and Outcomes." SocArXiv, September 20, 2023. https://doi.org/10.31235/osf.io/zfnuh.

Nettle, Daniel, Zoe Harper, Adam Kidson, Rosie Stone, Ian S. Penton-Voak, and Melissa Bateson. "The Watching Eyes Effect in the Dictator Game: It's Not How Much You Give, It's Being Seen to Give Something." *Evolution and Human Behavior* 34, no. 1 (January 1, 2013): 35–40. https://doi.org/10.1016/j.evolhumbehav.2012.08.004.

Noë, Ronald, and Peter Hammerstein. "Biological Markets." *Trends in Ecology & Evolution* 10, no. 8 (August 1, 1995): 336–39. https://doi.org/10.1016/S0169-5347(00)89123-5.

Nowak, Martin A., and Karl Sigmund. "Evolution of Indirect Reciprocity." *Nature* 437, no. 7063 (October 2005): 1291–98. https://doi.org/10.1038/nature04131.

"OECD Web Archive." https://web-archive.oecd.org/2021-05-11/588040-inheritance-estate-and-gift-taxes-could-play-a-stronger-role-in-addressing-inequality-and-improving-public-finances.htm, accessed November 29, 2023.

Orians, Gordon H. "On the Evolution of Mating Systems in Birds and Mammals." *The American Naturalist* 103, no. 934 (1969): 589–603.

Ostrom, Elinor. "Common-Pool Resources and Institutions: Toward a Revised Theory." In *Handbook of Agricultural Economics*, 2:1315–39. Agriculture and Its External Linkages. Elsevier, 2002. https://doi.org/10.1016/S1574-0072(02)10006-5.

Ostrom, Elinor. "Collective Action and the Evolution of Social Norms." *Journal of Economic Perspectives* 14, no. 3 (September 2000): 137–58. https://doi.org/10.1257/jep.14.3.137.

Ostrom, Elinor, and Roy Gardner. "Coping with Asymmetries in the Commons: Self-Governing Irrigation Systems Can Work." *The Journal of Economic Perspectives* 7, no. 4 (1993): 93–112.

Oxoby, Robert J., and John Spraggon. "Mine and Yours: Property Rights in Dictator Games." *Journal of Economic Behavior & Organization* 65, no. 3 (March 1, 2008): 703–13. https://doi.org/10.1016/j.jebo.2005.12.006.

Parfit, Derek. *On What Matters: Volume Two*. The Berkeley Tanner Lectures. Oxford: Oxford University Press, 2011.

Phillips, Tim, Jiawei Li, and Graham Kendall. "The Effects of Extra-Somatic Weapons on the Evolution of Human Cooperation Towards Non-Kin." *PLOS ONE* 9, no. 5 (May 5, 2014): e95742. https://doi.org/10.1371/journal.pone.0095742.

Pierotti, Raymond. "Gull-Puffin Interactions on Great Island, Newfoundland."

Biological Conservation 26, no. 1 (May 1, 1983): 1–14. https://doi.org/10.1016/0006-3207(83)90044-7.

Pizzol, Damiano, Mike Trott, Laurie Butler, Yvonne Barnett, Tamsin Ford, Sharon A. S. Neufeld, Anya Ragnhildstveit, et al. "Relationship Between Severe Mental Illness and Physical Multimorbidity: A Meta-Analysis and Call for Action." *BMJ Mental Health* 26, no. 1 (October 1, 2023). https://doi.org/10.1136/bmjment-2023-300870.

Rabinowitch, Tal-Chen, and Andrew N. Meltzoff. "Synchronized Movement Experience Enhances Peer Cooperation in Preschool Children." *Journal of Experimental Child Psychology* 160 (August 1, 2017): 21–32. https://doi.org/10.1016/j.jecp.2017.03.001.

Raffa, Kenneth F., Erinn N. Powell, and Philip A. Townsend. "Temperature-Driven Range Expansion of an Irruptive Insect Heightened by Weakly Coevolved Plant Defenses." *Proceedings of the National Academy of Sciences* 110, no. 6 (February 5, 2013): 2193–98. https://doi.org/10.1073/pnas.1216666110.

Rathore, Akanksha, Kavita Isvaran, and Vishwesha Guttal. "Lekking as Collective Behaviour." *Philosophical Transactions of the Royal Society B: Biological Sciences* 378, no. 1874 (February 20, 2023): 20220066. https://doi.org/10.1098/rstb.2022.0066.

Reby, David, and Karen McComb. "Anatomical Constraints Generate Honesty: Acoustic Cues to Age and Weight in the Roars of Red Deer Stags." *Animal Behaviour* 65, no. 3 (March 1, 2003): 519–30. https://doi.org/10.1006/anbe.2003.2078.

Ronson, Jon. *The Psychopath Test: A Journey Through the Madness Industry*. New York: Riverhead, 2011.

Roozenbeek, Jon, Sander van der Linden, Beth Goldberg, Steve Rathje, and Stephan Lewandowsky. "Psychological Inoculation Improves Resilience Against Misinformation on Social Media." *Science Advances* 8, no. 34 (August 24, 2022): eabo6254. https://doi.org/10.1126/sciadv.abo6254.

Rousseau, Jean-Jacques. *Discourse on the Origin and Foundations of Inequality Among Men: By Jean-Jacques Rousseau with Related Documents*. London: Palgrave, 2011.

Schmid, Monika, and Peter L. Patrick. "The Trouble with Relying on How People Speak to Determine Asylum Cases." *The Conversation*, March 12, 2015. http://theconversation.com/the-trouble-with-relying-on-how-people-speak-to-determine-asylum-cases-38562.

Schultheiss, Patrick, Sabine S. Nooten, Runxi Wang, Mark K. L. Wong, François Brassard, and Benoit Guénard. "The Abundance, Biomass, and Distribution of Ants on Earth." *Proceedings of the National Academy of*

Sciences 119, no. 40 (October 4, 2022): e2201550119. https://doi.org/10.1073/pnas.2201550119.

Sen, Amartya. *The Idea of Justice.* Cambridge: Harvard University Press, 2009. https://doi.org/10.2307/j.ctvjnrv7n.

Simpson, George Gaylord. *Tempo and Mode in Evolution.* New York: Columbia University Press, 1944.

Singer, Peter. *The Expanding Circle.* Princeton: Princeton University Press, 2011. https://press.princeton.edu/books/paperback/9780691150697/the-expanding-circle.

Singh, Manvir. "How Dowries Are Fuelling a Femicide Epidemic." *New Yorker,* June 12, 2023. https://www.newyorker.com/magazine/2023/06/19/how-dowries-are-fuelling-a-femicide-epidemic.

Singh, Manvir. "The Idea of Primitive Communism Is as Seductive as It Is Wrong | Aeon Essays." https://aeon.co/essays/the-idea-of-primitive-communism-is-as-seductive-as-it-is-wrong, accessed November 21, 2023.

Singh, Manvir. "Subjective Selection and the Evolution of Complex Culture." *Evolutionary Anthropology: Issues, News, and Reviews* 31, no. 6 (2022): 266–80. https://doi.org/10.1002/evan.21948.

Singh, Manvir, and Luke Glowacki. "Human Social Organization During the Late Pleistocene: Beyond the Nomadic-Egalitarian Model." *Evolution and Human Behavior* 43, no. 5 (September 1, 2022): 418–31. https://doi.org/10.1016/j.evolhumbehav.2022.07.003.

Singh, Manvir, and Joseph Henrich. "Why Do Religious Leaders Observe Costly Prohibitions? Examining Taboos on Mentawai Shamans." *Evolutionary Human Sciences* 2 (January 2020): e32. https://doi.org/10.1017/ehs.2020.32.

Singh, Manvir, and Moshe Hoffman. "Commitment and Impersonation: A Reputation-Based Theory of Principled Behavior." PsyArXiv, January 13, 2021. https://doi.org/10.31234/osf.io/ua57r.

Singhal, Atul. "Long-Term Adverse Effects of Early Growth Acceleration or Catch-Up Growth." *Annals of Nutrition & Metabolism* 70, no. 3 (2017): 236–40. https://doi.org/10.1159/000464302.

Smith, Adam. *The Wealth of Nations.* CreateSpace Independent Publishing Platform, 2014; originally published in 1776.

Smith, Eric Alden, Rebecca Bliege Bird, and Douglas W. Bird. "The Benefits of Costly Signaling: Meriam Turtle Hunters." *Behavioral Ecology* 14, no. 1 (January 1, 2003): 116–26. https://doi.org/10.1093/beheco/14.1.116.

Sogbanmu, Temitope Olawunmi, Heather Sauyaq Jean Gordon, Lahcen El Youssfi, Fridah Dermmillah Obare, Seira Duncan, Marion Hicks, Khadeejah Ibraheem Bello, Faris Ridzuan, and Adeyemi Oladapo Aremu. "Indigenous Youth Must Be at the Forefront of Climate Diplomacy." *Nature* 620, no. 7973 (August 2023): 273–76. https://doi.org/10.1038/d41586-023-02480-1.

Southern Poverty Law Center. "Changing of the Guard." Intelligence Report, August 29, 2001 (Fall 2001 issue). https://www.splcenter.org/fighting-hate/intelligence-report/2001/changing-guard, accessed November 21, 2023.

Spiegelhalter, David. *The Art of Statistics: Learning from Data*. UK USA Canada Ireland Australia India New Zealand South Africa: Pelican, 2019.

Starkweather, Katherine E., and Raymond Hames. "A Survey of Non-Classical Polyandry." *Human Nature* 23, no. 2 (June 1, 2012): 149–72. https://doi.org/10.1007/s12110-012-9144-x.

Stein, Alexandra. "How Cult Leaders Brainwash Followers for Total Control | Aeon Essays." https://aeon.co/essays/how-cult-leaders-brainwash-followers-for-total-control, accessed November 21, 2023.

Stewart, Tamara. "Investigating the Calusa." *Florida Museum*, September 25, 2020. https://www.floridamuseum.ufl.edu/science/investigating-the-calusa/.

Stibbard-Hawkes, D.N.E., R. D. Attenborough, and F. W. Marlowe. "A Noisy Signal: To What Extent Are Hadza Hunting Reputations Predictive of Actual Hunting Skills?" *Evolution and Human Behavior*, 2018. https://doi.org/10.1016/j.evolhumbehav.2018.06.005.

Stibbard-Hawkes, Duncan N. E. *Egalitarianism and Democratized Access to Lethal Weaponry: A Neglected Approach*. McDonald Institute for Archaeological Research, 2020. https://www.repository.cam.ac.uk/handle/1810/313537.

Strassmann, Beverly I. "The Function of Menstrual Taboos Among the Dogon." *Human Nature* 3, no. 2 (June 1, 1992): 89–131. https://doi.org/10.1007/BF02692249.

Sulak, M., L. Fong, and K. Mika. "TP53 Copy Number Expansion Is Associated with the Evolution of Increased Body Size and an Enhanced DNA Damage Response in Elephants." *eLife* 5:e11994 (2016).

Supran, Geoffrey, and Naomi Oreskes. "Assessing ExxonMobil's Climate Change Communications (1977–2014)." *Environmental Research Letters* 12, no. 8 (August 2017): 084019. https://doi.org/10.1088/1748-9326/aa815f.

Swami, Viren, Adrian Furnham, Nereshnee Balakumar, Candy Williams, Kate Canaway, and Debbi Stanistreet. "Factors Influencing Preferences for Height: A Replication and Extension." *Personality and Individual Differences* 45, no. 5 (October 1, 2008): 395–400. https://doi.org/10.1016/j.paid.2008.05.012.

Swift, Jonathan. *Tale of a Tub*. Edited by Kathleen Williams. London: Everyman Paperback Classics, 1975; originally published in 1704.

Bibliography

Tett, Gillian. *Anthro-Vision: How Anthropology Can Explain Business and Life*. London: Random House Business, 2021.

Trivers, Robert L. "The Evolution of Reciprocal Altruism." *Quarterly Review of Biology* 46, no. 1 (1971): 35–57.

Tullock, Gordon. *Social Dilemma: Of Autocracy, Revolution, Coup d'Etat, and War*, volume 8. Indianapolis: Liberty Fund, 2004.

Tunçgenç, Bahar, and Emma Cohen. "Interpersonal Movement Synchrony Facilitates Pro-Social Behavior in Children's Peer-Play." *Developmental Science* 21, no. 1 (2018): e12505. https://doi.org/10.1111/desc.12505.

Venkataraman, Vivek V., Thomas S. Kraft, and Nathaniel J. Dominy. "Tree Climbing and Human Evolution." *Proceedings of the National Academy of Sciences* 110, no. 4 (January 22, 2013): 1237–42. https://doi.org/10.1073/pnas.1208717110.

"Violence: Our Deadly Epidemic and Its Causes | Office of Justice Programs." https://www.ojp.gov/ncjrs/virtual-library/abstracts/violence-our-deadly-epidemic-and-its-causes, accessed November 29, 2023.

Vollrath, Fritz. "Uncoupling Elephant TP53 and Cancer." *Trends in Ecology & Evolution* 38, no. 8 (August 2023): 705–7. https://doi.org/10.1016/j.tree.2023.05.011.

Walker, Robert S., Mark V. Flinn, and Kim R. Hill. "Evolutionary History of Partible Paternity in Lowland South America." *Proceedings of the National Academy of Sciences* 107, no. 45 (November 9, 2010): 19195–200. https://doi.org/10.1073/pnas.1002598107.

Wallace, Elaine, Isabel Buil, and Leslie de Chernatony. "'Consuming Good' on Social Media: What Can Conspicuous Virtue Signalling on Facebook Tell Us About Prosocial and Unethical Intentions?" *Journal of Business Ethics* 162, no. 3 (March 1, 2020): 577–92. https://doi.org/10.1007/s10551-018-3999-7.

Wells, Jonathan C. K. "An Evolutionary Perspective on Social Inequality and Health Disparities: Insights from the Producer–Scrounger Game." *Evolution, Medicine, and Public Health* 11, no. 1 (January 1, 2023): 294–308. https://doi.org/10.1093/emph/eoad026.

Wells, Jonathan C. K. *The Metabolic Ghetto: An Evolutionary Perspective on Nutrition, Power Relations, and Chronic Disease*. Illustrated edition. Cambridge: Cambridge University Press, 2016.

West, S. A., A. S. Griffin, and A. Gardner. "Social Semantics: Altruism, Cooperation, Mutualism, Strong Reciprocity, and Group Selection." *Journal of Evolutionary Biology* 20, no. 2 (March 2007): 415–32. https://doi.org/10.1111/j.1420-9101.2006.01258.x.

West-Eberhard, Mary Jane. "Sexual Selection, Social Competition, and Speciation." *The Quarterly Review of Biology* 58, no. 2 (June 1983): 155–83. https://doi.org/10.1086/413215.

Whately, Richard (archbishop of Dublin). *Detached Thoughts and Apophthegms Extracted from Some of the Writings of Archbishop Whately.* Blackader, 1854.

Wiessner, Polly. "Experimental Games and Games of Life Among the Ju/'hoan Bushmen." *Current Anthropology* 50, no. 1 (February 2009): 133–38. https://doi.org/10.1086/595622.

Wiessner, Polly. "Risk, Reciprocity, and Social Influences on !Kung San Economics." In *Politics and History in Band Societies*, ed. Eleanor Leacock and Richard Lee, 61–84. Cambridge: Cambridge University Press, 1982.

Wilkinson, Richard G. "Income Inequality, Social Cohesion, and Health: Clarifying the Theory—a Reply to Muntaner and Lynch." *International Journal of Health Services* 29, no. 3 (1999): 525–43.

Williams, Bernard. *Essays and Reviews: 1959–2002.* Princeton: Princeton University Press, 2015. https://press.princeton.edu/books/paperback/9780691168609/essays-and-reviews.

Winegard, Bo, Ben Winegard, and David C. Geary. "The Status Competition Model of Cultural Production." *Evolutionary Psychological Science* 4, no. 4 (December 1, 2018): 351–71. https://doi.org/10.1007/s40806-018-0147-7.

Wrangham, Richard W. "Targeted Conspiratorial Killing, Human Self-Domestication, and the Evolution of Groupishness." *Evolutionary Human Sciences* 3 (2021): e26. https://doi.org/10.1017/ehs.2021.20.

Yamagishi, Toshio. "Exit from the Group as an Individualistic Solution to the Free Rider Problem in the United States and Japan." *Journal of Experimental Social Psychology* 24, no. 6 (November 1, 1988): 530–42. https://doi.org/10.1016/0022-1031(88)90051-0.

Yamagishi, Toshio. "The Provision of a Sanctioning System in the United States and Japan." *Social Psychology Quarterly* 51, no. 3 (1988): 265–71. https://doi.org/10.2307/2786924.

Yuan, Mingliang, Giuliana Spadaro, Shuxian Jin, Junhui Wu, Yu Kou, Paul A. M. Van Lange, and Daniel Balliet. "Did Cooperation Among Strangers Decline in the United States? A Cross-Temporal Meta-Analysis of Social Dilemmas (1956–2017)." *Psychological Bulletin* 148, no. 3–4 (2022): 129–57. https://doi.org/10.1037/bul0000363.

Zahavi, Amotz. "Mate Selection: A Selection for a Handicap." *Journal of Theoretical Biology* 53, no. 1 (September 1, 1975): 205–14. https://doi.org/10.1016/0022-5193(75)90111-3.

Zerjal, Tatiana, Yali Xue, Giorgio Bertorelle, R. Spencer Wells, Weidong Bao, Suling Zhu, Raheel Qamar, et al. "The Genetic Legacy of the Mongols." *American Journal of Human Genetics* 72, no. 3 (March 2003): 717–21.

Acknowledgments

Another of Arthur Machen's statements that stands out to me is that writers dream in fire but work in clay. At many moments when writing this book I have felt that I can't find the fire that drove me to start it, and consequently my words have often not conveyed the thoughts that preceded them. I am lucky to have many people around me who have provided the fire when my own has failed, and without whom this book wouldn't be anything approaching a work with which I am content. These are, in no particular order, Nik Chaudhary, Daniel Nettle, Robert Attenborough, Carol Brayne, Devi Sridhar, Athena Aktipis, Jon Ronson, Manvir Singh, Kate Douglas, Jonathan Weigel, Jo Wolff, Elizabeth Sylvia, Gillian Tett, Danielle Sweeney, Sen Bhuvanendra, Steve Duffy, Will Millard, Phillip King, and Laura Van Holstein. I am especially thankful to my doctoral adviser, Robert Foley, my friend and mentor David Lahti, my former academic adviser Anthony Price, my close friend Amy St. Johnston, my agent, Emma Bal, my editor at Yale University Press, Jean Thomson Black, Adam LeBor, and Nick Humphrey, all of whom have given me critical guidance and feedback throughout this project, which, while sometimes difficult to hear, was invaluable. My mother, Wendy Orent, has given me similarly invaluable feedback, on both my writing and my thinking; I am also grateful to my late father-in-law, Merv Lebor, and my mother-in-law, Jo Lebor, for reasons, including the obvious ones, that extend past anything to do with this book. *Invisible Rivals* is dedicated to my wife, Daniella Lebor, our son, Misha Lev, and my late father, Mitchell Goodman.

Index

accents, in speech, 80–81, 95
acculturation, 36–37, 89, 102
Acemoglu, Daron, 187n41
Aché people, 23–25, 33, 40, 41, 111
activists, overnight, 189n34
adaptations, 35–37, 98–100
adaptive therapy, 177
Adidas, 79
adultery, 96–99, 193nn52 and 54
age, and circles of duty, 153–54
aggression: inequality in relation to, 151–53; proactive, 41–42, 67, 189n38; reactive, 21–22, 62, 67, 189n29
agricultural societies, 87–88, 192n38
agriculture: common pool resources in, 141–42; Lysenkoism in, 8–10
Aktipis, Athena, 24, 49
Alexander, Richard, 56, 80
Alliance of Motion Picture and Television Producers, 139
alphas (dominants): human, 21–22, 41, 62–63, 165; non-human, 5, 45
altruism: binary of selfishness and, xiv, 7–8; capital maximization and, 90–91; in charitable giving, 76–78, 91; definitions of, 76, 78, 183; false, 76–78, 88–90; mathematical model of, 1; reciprocal, 50, 52, 55; sacrifice in, 88, 91, 132; skepticism about existence of, xiii, 90–91; in social capital, 78–79; true, 90–91, 106–7
American Humanist Association, 122
anarchists, 17–20, 29
anarcho-communism, 19–20
ancient societies: cultural diversity of, 16, 33; deception and exploitation in, 16–17, 29–30; egalitarianism in, 21–23, 29, 31–32, 39–40; noble savages in, xiii, 17, 21–22, 39–40; tools in, 15–16, 21–23
animals: cancer in, 46–47; coevolutionary arms race in, 92–93; competition in, 5, 18; deception in, 4–5, 45–46; embodied capital in, 81–83; environmental influences on behaviors of, 86–87; mutual aid in, 18–20, 50, 52; scrounging by,

animals (*continued*)
  151–52; signals used by, 58–59, 82–83
Aniston, Jennifer, 126
Anna Karenina (Tolstoy), 64
anonymity: in charitable giving, 78, 81, 91; in economic games, 29, 112, 114–15; in invisibility of deception, 96, 106–7; in small vs. large societies, 41–42, 96, 107; in social media, 10
Anscombe, Elizabeth, x
anthropology, views of human nature in, viii–ix, 20
Anthro-vision (Tett), 175
anti-Semitism, 13, 79, 160
ants, 34–36
Aristotle, 89, 101, 194n22, 200n22
artificial intelligence (AI), 138–39
asylum seekers, 95
atheism, 122–24
Atran, Scott, 159–60
attraction, basin of, 138
Australian Aboriginals, 31, 37
autocracies, 63, 135, 137–38, 167
Axelrod, Robert, 50, 54

Baby Keem, 189n34
Babylonians, 100
bacteria, 5
banking, Swiss, 120–21
Bankman-Fried, Sam, 76
Barker, Alison, 95
Barshai, Toman, 86–87
basin of attraction, 138
bats, 52
BaYaka people, 27
beetles, pine, 92
Bentham, Jeremy, x
Bezos, Jeff, 76, 89, 168
bias: in ethnographic research, 20; against homelessness, 147–48; in moral licensing, 116–17; self-deception about, 172–74; toward conformity and prestige, 135–38; white hat, 172
billionaires, 76–78, 89, 149, 155–56
biological altruism, definition of, 76, 183. *See also* altruism
biological cooperation. *See* cooperation
biological selfishness, definition of, 2, 184. *See also* selfishness
Biology of Moral Systems, The (Alexander), 56
birds: brood parasitism in, 5, 93, 96, 152; embodied capital in, 82–83; parasite resistance in, 58–59, 82, 157; resource capital in, 75; scrounging in, 152
Bitcoin, 127
blindness, to consequences in natural selection, 47–49
blind trust, 166
Blind Watchmaker, The (Dawkins), 47–48
Branch Davidians, 14
Brayne, Carol, 171
Brazil, fisheries in, 143
Breaking Bad (TV show), 104
bridewealth, 87
Brookings Institution, 150
Buddhists, 80
burqas, 123
Bush, George H. W., 106
businesses: deception about ethical practices in, 4; influence on policymaking, 158; psychopaths in positions of power in, 3, 128, 137; sustainability practices of, 160–62, 180
Byrne, Richard, 45

Index

Calusa, 33–34, 37
cancel culture, 10
cancer: adaptive therapy for, 177; biological mechanisms for containing, 91–92; blindness to consequences in, 47–49; capital maximization compared to, 90; childhood, 125–26; deception in, 5, 8, 45, 49, 164; free riders compared to, 8, 46–49, 165; health inequality in, 149–50; hosts killed by, 47–49, 69, 90, 165–66; invisible rivals compared to, 69, 165; prehistoric existence of, 46–47
capital, 71–103; cultural and physical context of, 86–88; defenses against maximization of, 91–103; definitions and use of term, 73–74; in punishments for norm-breaking, 99; sacrifice of, 88, 91, 132, 157–59; universal drive to maximize, 88–91, 128
capital, intergenerational transmission of, 74; in development of inequality, 87–88, 155–56; embodied capital, 88; resource capital, 78, 87–88, 155–56; social capital, 79–81, 87–88; taxation on, 155–56
capitalism, 74, 79, 88, 102, 114, 151
capital punishment, 99
carbon neutrality, 160–61
cash transfers, unconditional, 147–48
Catholicism, 122
Cave, allegory of the, 162–63
celibacy, involuntary, 99
cell death, programmed, 92
censorship, soft, 9–10
Center for the Study of Violence, 153

centrarchid sunfish, 92–93
charismatic leaders, 14
charitable giving: anonymity in, 78, 81, 91; by billionaires, 76–78, 155; to St. Jude Hospital, 125–26
Charity Watch, 126
ChatGPT, 138–39
Chaudhary, Nikhil, 27–29
cheating: definition of, 183; rationalizing about, 117; reputational effects of, 57; in small vs. large societies, xi, 96; thriving through, 48–49, 168; unknown prevalence of, 106–7. *See also* deception; free riders
Cheating Cell, The (Aktipis), 49
chemical defenses, 92
Cheyenne people, 30
children: cancer in, 125–26; costs of rearing, 77; duration of childhood, 86, 136; embodied capital in, 85–86; malnutrition in, 149–50; paternity of, 96–99, 193nn52 and 54
chimpanzees: cultural transmission in, 136; deception in, 5, 45–46, 69, 164; duration of childhood in, 86; reactive aggression in, 189n29; tool use by, 21
Christian Identity movement, 13
Christianity, 122, 190n2
circles of duty, 148–56, 180
Claridge's hotel (London), 148–49
climate change: carbon offsetting in, 160–62; coevolutionary arms race in, 92; free riders in, 146, 178–79; indigenous populations in, 175–76; motivations in, 172
Clinton, Hillary, 125
cocaine market, 120

CoDa (Cooperation Databank), 109–10
Code of Hammurabi, 100
Codes of the Underworld (Gambetta), 174
coercive control, 30, 32–33
coevolutionary arms race, 92–93
color badges, 58, 157
common pool resources, 141–44
commons, tragedy of the, 140–44
communication. *See* language
communism, 8–9, 19–20
competition: in animals, 5, 18; binary of cooperation and, xiv, 6–8, 171; in supercompetitors, 70; universality of, 5, 165; as visible rivalry, x. *See also* exploitation
con artists, 12–13, 15
conformity, 132–38
conjoint surveys, 154–55
conservatism, 153–54
contracting circles of duty, 148–56
cooperation, 49–70; binary of competition and, xiv, 6–8, 171; as core of human nature, 16, 130, 165–66, 181; cultural group selection in, 62–63; definitions of, 16, 183, 186n6; education about importance of, 162; with exploiters, 42; intelligence in, 55–57, 67; intention in, 65–69; kinship in, 49–52, 60; language in, 56–62; in non-human species, 5, 51–52; problem of opportunity in, 65–70, 114; as reason for human success, 5, 16, 68; reciprocity in, 50, 52–56, 60, 72; reputation in, 55–57, 60; self-interest as compatible with, 63; social selection in, 60–61; spread of, in evolutionary history, xiii–xiv; in supercooperators, 6, 68, 70; survival of the friendliest in, 61–63
Cooperation Databank (CoDa), 109–10
corporations. *See* businesses
corruption: of goodness, 104–5; government, 19–20; invisibility in, 44; in religion, 122. *See also* exploitation
costly signaling hypothesis, 85
costs: of child rearing, 77; in definition of altruism, 76; of earning trust, 174; of signaling, 58–60, 65, 76, 85, 157–59
Covid-19 pandemic, 125, 145, 149, 158–59
creativity, in exploitation, 102, 128
credibility, 167–70
Credit Suisse, 121
criminal gangs, 174
critical thinking, 170–72
crypsis, 126
cryptocurrencies, 127
cuckolds, 193n52
cuckoos, 5, 93, 94, 96, 152, 193n52
cult leaders, 12–15, 137
cultural adaptations, 36–37, 98–100
cultural diversity, ancient, 16, 33
cultural evolution, subjective, 102
cultural evolutionary studies, 135–37
cultural groups: cultural group selection, 62–63; definition of, xiii; signs of membership in, 95. *See also* societies
cultural immune system, 99–100, 102

Index

cultural norms. *See* norms
culture(s): acculturation into, 36–37, 89, 102; capital in context of, 86–88; as distinguishing feature of humans, 15–16; economic games in context of, 110–16; environmental influences on, 86–87; in escape from evolution, xiii–xiv; of inequality, 151–56; prosociality in context of, 111–13; transmission of, 36, 135–38
Curb Your Enthusiasm (TV show), 78, 91
cyclones, 25

Danson, Ted, 78, 91
D.A.R.E., 106
Darwin, Charles: *The Descent of Man,* 5, 77, 82; misunderstandings of, 18–19; *The Origin of Species,* xiii, 18; on sexual selection, 82, 83; on social instinct, 5, 164. *See also* natural selection
David, Larry, 78, 91
David Copperfield (Dickens), viii
Dawkins, Richard, 47–48, 123–24
death penalty, 99
deception: in ancient societies, 16–17; diversity of opportunities for, 46; in European vs. non-European societies, 16–17; through language, 45, 46, 67–68, 164; through mimicry, 42, 45, 93; in non-human species, 4–5, 45–46; punishment for, 60, 61–62; rates of detection of, 60, 167–68; rewards for, 48–49, 168–69; in signals, 64–66; in small vs. large societies, xi, 41–42, 96, 107; universality of,

5, 49; unknown prevalence of, 68, 106–7; ways of avoiding detection, 46. *See also* self-deception
deer, red, 59
defeso policy, 143
dementia, 171
democracies, 10, 17, 63, 74, 138, 179
Democratic party, U.S., 125
demographic transition, 85–86
Dennett, Daniel, 196n40
Descartes, René, viii
Descent of Man, The (Darwin), 5, 77, 82
despotism, 165, 166
devout actors, 66, 173
de Waal, Frans, 107
diabetes, 150
Dickens, Charles, viii
dictator games, 114–15, 117
dictatorships, 137–38
dilemmas: Prisoner's, 52–55, 64; social, 113, 141–44
direct fitness, 183
disease transmission, 99–100
disinformation, 156, 171
Dogon people, 96
dogs, 21, 62
domestication, self-, 8, 21, 62, 67
dominants. *See* alphas
Downing, James R., 126
dowries, 87
drinking water, 141–42
drug use, 120, 147
dualism, viii
Dunbar, Robin, 55
Dunbar's number, 55, 57
duty, circles of, 148–56, 180
Dwyer, Ryan, 108, 109
dyadic games, 117
dyadic relationships, 72

economic games, 110–17; anonymity in, 29, 112, 114–15; dictator, 114–15, 117; dyadic, 117; prosociality vs. selfishness in, 110–16; surveillance in, 113–15, 194n14
economic inequality. *See* resource inequality
economic selfishness, 6
education: about ethical and critical thinking, 170–72, 180–81; about importance of cooperation, 162; in life expectancies, 150–51; religion in, 122
Effective Altruism, 76
egalitarianism: in ancient societies, 21–23, 29, 31–32, 39–40; definitions of, 31–32; in hunter-gatherer groups, 23, 27–34, 40–41; modes of exploitation and, 37–39; problems with myth of, 42; U-shaped history of, 21–22
elders, in gerontocracies, 30–31, 33, 39
elections, U.S. presidential, 125, 137
electricity, 139–40, 142, 162–63
"Electrification" (Zoshchenko), 131, 140, 162
elephants, 92
Elohim City, 13–14
embodied capital, 81–86; intergenerational transmission of, 88; maximization of, 89; in non-human species, 81–83; in punishments for norm-breaking, 99; in reproductive success, 82–86
Emerson, Ralph Waldo, 153
enforcement: of norms, in small vs. large societies, 96; in solving of social dilemmas, 143. *See also* punishment
Engels, Friedrich, 18
environments: capital in context of, 86–88; influence on human and animal behaviors, 86–87; specialization in exploitation of, 34–37; sustainability practices and, 146, 160–62, 180
Ephraimites, 94
equality. *See* inequality/equality
Erasmus, 74
ethics and morality: in business, deception about, 4; corruption of goodness in, 104–5; education about, 170–71, 180–81; expanding vs. contracting circles in, 148–56, 180; Golden Rule in, 72–73; market forces in, 118–21; moral licensing in, 116–17; in reciprocity norms, 101; in Trolley Problem, 54; in virtue signaling, 64–66
ethnographic research, 20, 23–34. *See also* hunter-gatherer groups
European Union, 117, 156
"Evolution of Cooperation, The" (Hamilton and Axelrod), 54
"Evolution of Reciprocal Altruism, The" (Trivers), 52
expanding circles of duty, 148–56, 180
exploitation: in ancient societies, 16–17, 29–30; of blind trust, 166; cooperation with, 42; by cult leaders, 12–15; cultures of inequality in, 152–56; inheritance of resource capital in, 88; innovation in methods of, 102, 128; by institutions, 122–29; mechanisms for containing, 91–95,

177; modes of, 37–40; and social norms, 15, 39, 66; social status in, 31, 38–39; specialization in, 34–37; as universal in human societies, 29–30, 40–41, 49, 69; as universal in natural world, 5, 164
extractive state, 187n41
ExxonMobil, 172
"eye for an eye," 100

facial expressions, 59
fairies, 181
fairness, 114–18; pretense of innate preference for, 42–43, 111, 114–16, 121; rationalizing about, 114–18, 120
false altruism, 76–78, 88–90
false indicators of trust, 157, 162, 174
false signals, 62, 64–66
fascism, 132, 134
Faux, Zeke, 127
Fehr, Ernest, 111–12, 120
feminism, 123–25
Financial Times (newspaper), 139, 148–49, 168
First Nations, 175
fisheries, 140, 142–43
fitness: direct, 183; inclusive, 50–51, 55, 183
Fitzgerald, F. Scott, 115
Foley, Robert A., vii–xi
food: hiding of, 45–46; in malnutrition, 149–51; scrounging for, 151–52; sharing of, 23–29, 33
foraging societies, 32–34
Forest, The (Zeller), 104
forest fires, 175
Foroohar, Rana, 139
freedom, personal, 164, 167

free riders, 46–49, 144–48; cancer as type of, 8, 46–49, 165; in climate change, 146, 178–79; definitions of, 8, 145, 183; homeless people as, 147–48; how to treat, 146, 177–81; invisibility of, 144; norms against, 66, 144–45; persistence of, 144–47, 165; punishment of, 62–63, 146–47, 179–81; strategies for thwarting, 8, 144, 146–47; taxation of, 144–47
Friedman, Milton, 9
friendliest, survival of the, 61–63, 67
friendosexuality, 14

Gallic War (Caesar), 98
Gambetta, Diego, 174
gangs, 174
Garfield, Zachary, 98–99
Gaza Strip, 159–60
Gelman, Audrey, 125
gender equality, 123–25
generosity research, 107–17
genetics: horizontal gene transfer in, 48; in kin selection, 50–51; in specialization, 35–37
genetic success, 51, 60, 76, 80
Genghis Khan, 77
Gen Z, 154
Germany, Nazi, 120–21, 131–33, 137
gerontocracies, 30–31, 33, 39
Gestapo, 131–32
Gift, The (Mauss), 71
gift giving, 71–73, 100–101, 102
giving. *See* charitable giving; economic games; gift giving
Giving Pledge, 77–78, 155
Glass, Stephen, 105–7
Global Carbon Project, 178

Global Indigenous Youth Summit on Climate Change, 176
Global Tax Evasion Report, 156
global warming. *See* climate change
Glowacki, Luke, 32–34
Goering, Albert, 131–32, 133, 163
Goering, Hermann, 131–32, 163
GoFundMe, 126
Golden Rule, 72–73, 101–2, 190n2
Goldman Sachs, 129
Gold Standard, 161
Good Natured (de Waal), 107
goodness, corruption of, 104–5
gossip, 56–57
government: corruption in, 19–20; libertarianism views on size of, 17; surveillance by, 113–14
Grafen, Alan, 50
Gravity Payments, 3–4
Great Gatsby, The (Fitzgerald), 115
greenwashing, 161–62, 198n49
Grevenbroek, Johannes Gulielmus de, 31
grooming, 52
Grossman, Vasily, 9
Guardian (newspaper), 4, 120, 161
Gucci company, 161
Guérewol festival, 83, 84

Hadza people, 25, 111
Haldane, J.B.S., 51
Hamas, 159–60
Hamilton, William, 1, 50–52, 54, 55
Hamilton's rule, 51, 52
Handicap Principle, 58–59
happiness, 164
Hardin, Garrett, 140
Hare, Brian, 21
Harris, Sam, 123–24
Harrison Bergeron (Vonnegut), 167

Harvard University, 153
health: inequality in, 149–51; market forces in, 129; public, 100, 158–59
Heidegger, Martin, 137
Helbrans, Shlomo, 14–15
Hendy-Freegard, Robert, 12–13, 15, 42
Henrich, Joseph, 38, 99, 110, 194n10
hierarchies, social, 7, 21, 37, 45, 58, 151–52
Hirst, Damien, 149
Hitchens, Christopher, 123–24
Hitler, Adolf, 131, 137
Hobbes, Thomas, viii, 108, 165
Holmberg, Allan, 31
Holocaust, 120–21, 131–32
homelessness, 147–48, 154
Homo economicus, 6–7, 116
Homo reciprocans, 6–7, 185n4
honey, 27
horizontal gene transfer, 48
horticultural societies, 87, 192n38
Hottentots. *See* Khoekhoen people
housing crises, 154
Humanists U.K., 122, 195n29
human nature, views of: anthropological, viii–ix, 20; oversimplification in, xiv; philosophical, viii–ix; in policymaking, xiv–xv; political trends in relation to, 6, 9; self-deception about, 165
humans: culture as distinguishing feature of, 15–16; as single species, 35–36; slave-trafficking of, 127
humans, success of: cooperation in, 5, 16, 68; cultural transmission in, 136; through lying, 168; professional, accents in, 80–81;

reception of signals in, 60–61; self-interest in, 49
Hume, David, 29, 73
humor, 124
Humphrey, Nicholas, 67, 136–38, 166, 188n44
hunter-gatherer groups, 23–34; definition of, 192n38; egalitarianism in, 23, 27–34, 40–41; embodied capital in, 84–85; exploitation of women in, 30–31; number of social contacts in, 55; reactive aggression in, 189n29; reproductive success in, 84–85; sharing in, 23–29, 33; types of wealth in, 87
Huxley, Thomas, 25

Igliniit Project, 175
imitation, 42, 58, 66
immigrants, 95
immune system: cancer and, 5, 45, 91–92; cultural, 99–100, 102
inclusive fitness: definition of, 50, 183; in history of cooperation, 50–51, 55
indigenous populations, 175–76
indirect reciprocity, 56
inequality/equality, 148–58; aggression in relation to, 151–53; arithmetic vs. geometric, 194n22; capital in development of, 87–88, 155–56; cultures of, 151–56; gender, 123–25; in health, 149–51; inequity aversion in, 111, 115–16; in pay, 117; reproductive, 78, 152; resource (economic), 76, 100, 110, 148–54; sacrifice needed to reduce, 158; in sharing by hunter-gatherers, 26–27; social, 150–54. *See also* egalitarianism; fairness

inequity aversion, 111, 115–16
infanticide, 40
Inflation Reduction Act (2023), 179
inheritance: in development of inequality, 88, 155; of resource capital, 75, 76, 87; taxation of, 155–56
injustice, identifying, 177
In Praise of Folly (Erasmus), 74
insects, 34–36, 82–83
institutions, exploitation by, 122–29
instrumental aggression. *See* proactive aggression
intelligence: artificial, 138–39; in history of cooperation, 55–57, 67; Machiavellian, 41–42, 46, 67, 188n44; as tool for exploitation and capital maximization, 128
intention: in altruism, 78; hiding of, 46, 58, 66–69, 89, 128, 130; problem of, in cooperation, 65–69
internalization, 101
International Brotherhood of Electrical Workers, 139
Internet access, 141–42
Inuit people, 175
invisibility: of capital, 88; in corruption, 44; of deception, anonymity in, 96, 106–7; of free riders, 144; through language, 46, 68; of trustworthiness, 157
involuntary celibacy, 99
irrigation, 141–42
Islam, 123, 135, 151
isolation, social, 41–42
Israel, 159–60

JAMA Oncology (journal), 46–47
Japan, economic games in, 113–14
jealousy, 97

Jewish people: aggression against, 151; anti-Semitism against, 13, 79, 160; Golden Rule of, 190n2; in Holocaust, 120–21, 131–33; paternity certainty and, 96–97; sexism and, 123
Johansson, Scarlett, 75–76
Jones, Ron, 133–35
Journal of Theoretical Biology, 52
Ju/hoansi (!Kung) people, 25, 40, 79, 111–12
Julius Caesar, 98, 182

Kant, Immanuel, x
Kapital, Das (Marx), 19
Kessler, Karl, 18–19, 186n11
Khanti people, 33
Khoekhoen people, 30–31, 187nn28–29
kindness, 17–19, 32–33
King Lear (Shakespeare), viii
kings, divine right of, 152
kinship: in history of cooperation, 49–52, 60; in hunter-gatherer groups, 23–24; in kin selection, 50–51, 60, 63; in nepotism, 23, 27, 63, 81; ways of detecting, 97
knowledge: in embodied capital, 85–86; through Internet access, 141; self-, in placement of trust, 169–70
Koresh, David, 14
Korman, Edward R., 120–21
KPMG UK, 161, 198n49
Kropotkin, Pyotr: on egalitarianism, 29, 40; vs. Hobbes, 108, 165; *Mutual Aid*, 18–20, 29, 30, 186n11
!Kung (Ju/hoansi) people, 25, 40, 79, 111–12
Kushner, Jared, 81

Lady Macbeth of the Mtsensk District (Leskov), 104
Lamalera, 79
Lamar, Kendrick, 189n34
Lamarck, Jean-Baptiste, 8
Lane, Charles, 106
language: accents in, 80–81, 95; deception through, 45, 46, 67–68, 164; emergence of, 57–58; in history of cooperation, 56–62; invisibility through, 46, 68; in judgment of others, 57, 59–61. *See also* signals
Language Assessment for Determination of Origin (LADO), 95
law, rule of, viii
leaders: cult, 12–15, 137; religious, 122
lekking, 83
Lenin, Vladimir, 139–40, 162
Leskov, Nikolai, 104
leukemia, 45
Leviathan (Hobbes), 165
Lev Tahor, 14–15
libertarianism, 17, 20, 29–30, 40
life expectancies, 150–51
Life and Fate (Grossman), 9
listening, 170–72, 175–77
literature: corruption of goodness in, 104–5; invisible rivals in, vii–viii
Lomekwi 3 site (Kenya), 21
Lud-in-the-Mist (Mirrlees), 182
Luftwaffe, 131
Lukas, Dieter, 86–87
lying: ability to detect, 60, 167–68; rewards for, 168–69; as unique to humans, 46. *See also* deception
Lysenko, Trofim, 8–9
Lysenkoism, 8–10, 20

Index

Maasai people, 24, 52, 79–80, 145
Macbeth (Shakespeare), 104–5
Machen, Arthur, 181
Machiavelli, Niccolò, 68
Machiavellian intelligence, 41–42, 46, 67, 188n44
mafia, 174
malnutrition, 149–51
Mandeville, Bernard, 49
Maori people, 71–72, 100–101, 102
marginalized groups, 176
market forces: in ethical decision-making, 118–21; in institutional exploitation, 129
marriage norms, 96–99
Marx, Karl, 10, 18, 19, 20, 36
Marxism, 8, 20, 74–75
Mating Game, The (documentary), 191n27
mating norms, 96–99
matrilineality, 87, 192n40
Mauss, Marcel, 71
Maynard Smith, John, 1–2
McCain, John, 168
men: in gerontocracies, 30–31, 33, 39; in hunter-gatherer groups, 23, 30–31; inequality and aggression in, 151; paternity of, 96–99, 193nn52 and 54; in patriarchies, 33; in patrilineality, 87, 192n40; sense of humor in, 124
Mendel, Gregor, 19
menstruation, 96–97, 193n54
Mentawai, 37–38
Menzel, Emil, 45–46
Meriam people, 84–85
mice, surplus, 118–19, 129, 194n24
mikvah, 96–97
Milgram, Stanley, 196n24
Mill, John Stuart, x
Millar, Robert G., 13–14

millennials, 154
mimicry, 42, 45, 93
mind-body dualism, viii
Mirrlees, Hope, 182
mirroring, 42
mole rats, naked, 95
monarchies, 152
money games. *See* economic games
monogamy, 23, 75–76, 99
Montaigne, Michel de, 16–17
Moore, John, 29–33, 38–39, 187n26
morality. *See* ethics and morality
moral licensing, 116–17
motivations, in placement of trust, 169–70, 172–74
Musk, Elon, 64, 76, 77, 89, 168
mutual aid, 18–20, 30, 50, 52. *See also* cooperation
Myers, P. Z., 124

narcissists, 105
Nash, John, 53
Nash equilibrium, 53
Native Americans, 30, 33–34
natural selection: blindness to consequences in, 47–49; reproductive success in, 19; in Russia, 18–19, 186n11; selfishness in, 8, 47–49; vs. social selection, 60
Nazi Germany, 120–21, 131–33, 137
need-based sharing, 23–24
negative norms, 101
neocortex, 55
Nepal, 142
nepotism, 23, 27, 63, 81
Netanyahu, Benjamin, 160
Netflix, 64
Nettle, Daniel, 154–55, 158
New Atheism, 122–24
Newman, Fred, 14
Newman Tendency, 14

New Republic, 105–6
news media, 158, 170, 180
New York Times, 65–66, 107–9
Nicomachean Ethics (Aristotle), 200n22
1984 (Orwell), 167, 181
noble savage, xiii, 17, 21–22, 39–40
Nolan, Francis, 81
nomadic-egalitarian model, 32
norms: breakers of, 96, 98–99, 130, 163, 165; and capital maximization, 96–103; choice in adoption of, 138; of cooperation, 9, 63; cultural transmission of, 36, 135–38; definition of, 193n59; in economic games, 113–14; enforcement in small vs. large societies, 96; evolution of, 142–44; and exploitation, 15, 39, 66; against free riders, 66, 144–45; functions of, 99, 101; internalization of, 101; libertarianism on, 17; of mating and marriage, 96–99; of reciprocity, 72–73, 101–2; redesigning of, challenges of, 142–44; self-interest in, 101–2; of sharing, 23–24
Nowak, Martin, 6
Nuer people, 23
nutrition, 149–51
Nyinba people, 97

Oates, Joyce Carol, 168–69
obesity, 149, 151
offsetting, carbon, 160–61
oleoresin, 92
oligarchies, 152
open societies, 137–38, 167, 181
Oppenheimer, Mark, 123–24
opportunity, for deception, 65–70; difficulty of resisting, 65, 105; ethical thinking and, 181; in humans vs. non-humans, 46; intention and, 65–69; in small vs. large societies, 41–42, 56, 96; surveillance and, 114
oppression, 30, 39
Organisation for Economic Co-operation and Development, 155
Organization, The (cult), 14
Origin of Species, The (Darwin), xiii, 18
Orthodox Judaism, 96–97
Orwell, George, 1–2, 167, 181
osotua, 24, 79–80
osteosarcoma, 47
ostracism, 57, 60, 62–63
Ostrom, Elinor, 140–43, 175
Othello (Shakespeare), vii–viii
others: language in judgment of, 57, 59–61; understanding of, in placement of trust, 169–70
overfishing, 140, 142–43
overnight activists, 189n34
overweight or obese people, 149

Palestinians, 159–60
Palmer, Walter, 57
Paradise Papers, 144
parasites: brood, 5, 93, 96, 152, 193n52; resistance to, 58–59, 82, 157
Parfit, Derek, 72, 190n5
Paris Agreement, 178–79
partible paternity, 97–98
partnerships, 60–61, 79
pastoralists, 37, 87–88, 192n38
paternity, 96–99, 193nn52 and 54
paternity certainty, 96–97
patriarchies, 33
patrilineality, 87, 192n40

Index

philanthropy. *See* charitable giving
philosophy, views of human nature in, viii–ix
pine beetles, 92
pine trees, 92
Plato, 44, 68, 162–63
Poliakov, Ivan, 18–19
policymaking: bottom-up vs. top-down, 137–43, 158–59; effects of lack of trust in, 156–62; how to earn trust in, 175–77; how to treat free riders in, 146, 177–81; implications of views of human nature in, xiv–xv; influence of the wealthy and businesses on, 158; universality of exploitation in, 41
politics: debates over human nature in relation to, 6, 9; effects of lack of trust in, 156–62; nepotism in, 63, 81; in scientific reasoning, 8–10
polyandry, 97–98
polygamy, 13–14, 38, 76
polygyny threshold, 75–76
Pondorfer, Andreas, 86–87
Popper, Karl, 196n40
poverty reduction, 154–55
Power of Darkness, The (Tolstoy), 105, 130
"Power of Darkness, The" (Singer), 105
power relations, 150–52
Pregnant Then Screwed, 77
prestige, bias toward, 135–38
Price, Dan, 4, 5, 7, 67
Price, George, 1–2, 7, 158
primates: deception in, 45–46, 164; intelligence of, 55; mutual aid in, 52; tool use by, 21
Prisoner's Dilemma, 52–55, 64

proactive aggression, 41–42, 67, 189n38
production, modes of, 34–37
professional success, speech accents and, 80–81
programmed cell death, 92
ProPublica, 64, 125–26
prosociality: cultural context of, 111–13; in economic games, 110–16; skepticism about, 2–6. *See also* altruism
psychological altruism, definition of, 78, 183. *See also* altruism
psychological selfishness, definition of, 184. *See also* selfishness
psychopaths: difficulty of identifying, 42, 105–6; embodied capital of, 89; motivations of, 174; in positions of power, 3, 128, 137; proactive aggression and intelligence in, 41–42; traits of, shared by everyone, 128–29
Psychopath Test, The (Ronson), 42
public health, 100, 158–59
public trust. *See* trust
punishment: for deception, 60, 61–62; for free riding, 62–63, 146–47, 179–81; for norm-breaking, 98–99; for selfishness, 8
punulu, 96
Putin, Vladimir, 89, 165

quarantine, 100
questioning, 170–73
quorum sensing, 5

Rand, Ayn, 6
rationalization, 116–21
rational man ideal, 116
rational selfishness, 110, 112, 115

Rawls, John, 194n21
reactive aggression, 21–22, 62, 67, 189n29
reciprocity: in cooperation, 50, 52–56, 60, 72; in gift giving, 71–72, 100–101; indirect, 56; norms of, 72–73, 101–2; in reciprocal altruism, 50, 52, 55
reinforcement, 168–69
relatedness, 51, 184
relational wealth, 27–29, 79–80
religion: atheists' critique of, 122–24; internalization of norms in, 101
reproductive and sexual success: of cult leaders, 13–14; embodied capital in, 82–86; inequality in, 78, 152; in natural selection, 19; resource capital in, 75–77; social status in, 38
Republic (Plato), 44, 68
Republican party, U.S., 170
reputation: in economic games, 112–13; in free-rider problem, 146–47, 180; in history of cooperation, 55–57, 60; and institutional exploitation, 126–27; language in, 56–57; in social crypsis, 126; in social selection, 79
resource(s): hunter-gatherers' sharing of, 23–29, 33; specialization in exploitation of, 34–37; in tragedy of the commons, 140–44
resource capital, 74–79; and charitable giving, 76–78; definition of, 74–75; exchange of, for social capital, 78–80, 91; intergenerational transmission of, 78, 87–88, 155–56; and life expectancies, 150–51; maximization of, 89; in punishments for norm-breaking, 99; in reproductive success, 75–77
resource (wealth) inequality: circles of duty and, 148–56; and generosity, 110; public trust and, 151, 156–57; reproductive patterns in, 76; taxation in reduction of, 100, 154–56; transmission of wealth in, 87–88, 155–56
reward-based experiments, 136
rewards, for deception, 48–49, 168–69
Rhine River, 73, 190n4
Rhone River, 73, 190n4
Ring of Gyges, 44, 68
risk pooling: definition of, 24; and free-rider problem, 144–45; in hunter-gatherer groups, 24–26, 41, 52
Robinson, James, 187n41
Ronson, Jon, 10, 42, 162, 168–69, 173–74
Rossel Islanders, 24–25
Rousseau, Jean-Jacques, viii, 17
Russia/Soviet Union: climate pledges by, 178–79; electricity regulation in, 139–40; invasion of Ukraine by, 64; Lysenkoism in, 8–10; natural selection in, 18–19, 186n11

Sacco, Justine, 10
sacrifice: in altruism, 88, 91, 132; in costs of signaling trust, 157–59
sage grouse, greater, 83
St. Jude Children's Research Hospital, 125–26
Schäublin, Cyril, 186n12
Schindler, Oskar, 132

Index

Schindler's List (film), 132
scientific reasoning, politics in, 8–10
scrounging, 151–52, 156
selection: cultural group, 62–63; kin, 50–51, 60, 63; against reactive aggression, 21; sexual, 8, 60, 82–83; social, 60–61, 79, 82. *See also* natural selection
self-censorship, 9–10
self-deception: about altruism, 90; about cooperative vs. exploitative human nature, 165; about motivations, 172–74; about trustworthiness, 157
self-domestication, 8, 21, 62, 67
self-interest, 6–8; cooperation as compatible with, 63; definition of, 184; maximization of, 8, 130; in norm design, 101–2; vs. selfishness, 63; in success of humans, 49
selfishness: binary of altrusim and, xiv, 7–8; cultural adaptations to control, 100; definitions of, 184; economic, 6; in economic games, 110–16; lack of trust based on, 166; masking of, in proactive aggression, 41–42, 67, 189n38; in natural selection, 8, 47–49; rational, 110, 112, 115; vs. self-interest, 63; universality of, 8–9
self-knowledge, in placement of trust, 169–70. *See also* self-deception
self-sacrifice. *See* sacrifice
Sen, Amartya, 177
sexism, 116–17, 123–24
sexual selection, 8, 60, 82–83
sexual success. *See* reproductive and sexual success
Shakespeare, William, vii–viii, 104–5
shamanism, 37–38
sharing: free riders and, 145; in hunter-gatherer groups, 23–29, 33
Shermer, Michael, 123–24
Shibboleth, 93–95
shock therapy experiments, 196n24
signals, 58–66; costs of, 58–60, 65, 76, 85, 157–59; deceptive, 62, 64–66; in history of cooperation, 58–66; in non-human species, 58–59, 82–83; reception of, in success of humans, 60–61; in sexual selection, 82–83; about social capital, 80–81; of trustworthiness, 61–62, 157–59; virtue, 64–66. *See also* language
Simpson, George Gaylord, 48
sincerity, 160
Singer, Isaac Bashevis, 105
Singer, Peter, 148
Singh, Manvir, 32–34, 38, 102
Sirionó people, 31
Skeptic (magazine), 123
Smith, Adam: on cooperation, 49–50; definition of capital by, 73; and economic games, 109–10, 112–13, 115; on rational man, 116
Smith, Alison, 124
Smoke (Turgenev), 12
social brain, 57, 66–67, 188n44
social capital, 78–81; definition of, 79; exchange of resource capital for, 78–80, 91; of free riders, 180; intergenerational transmission of, 79–81, 87–88; maximization of, 89; in punishments for norm-breaking, 99. *See also* reputation

social classes, 34, 80, 86–87
social contacts, number of, 55–56
social crypsis, 126
social dilemmas, 113, 141–44
social equality. *See* egalitarianism
social hierarchies, 7, 21, 37, 45, 58, 151–52
social inequality, 150–54
social instinct, 5, 164
social intelligence, 41–42, 188n44
socialism, 20, 74
social isolation, 41–42
social market, 60–61
social media: cancel culture in, 10; trustworthiness in, 168–69, 171–72
social norms. *See* norms
social selection, 60–61, 79, 82
social status and standing: altruism in, 78; badges of, 58, 157; in exploitation, 31, 38–39; in reproductive success, 38. *See also* social capital
social welfare programs, free-rider problem in, 145, 178, 179
societies: deception in small vs. large, xi, 41–42, 96, 107; exploitation as universal in, 29–30, 40–41, 49, 69
sociobiology, 2
sociopaths, 105
soft censorship, 9–10
Southern Poverty Law Center, 13–14
Soviet Union. *See* Russia/Soviet Union
specialization, 34–37
Spielberg, Steven, 132
Sridhar, Devi, 152, 158, 169
status. *See* social status and standing

stealing food, 151–52
Stein, Alexandra, 14
stone tools. *See* tools
strangers, cooperation with, 49–50, 52, 55–57, 60
subjective cultural evolution, 102
success: genetic, 51, 60, 76, 80; of individuals vs. species, 18–19, 49. *See also* humans, success of; reproductive and sexual success
sunfish, centrarchid, 92–93
supercompetitors, 70
supercooperators, 6, 68, 70
surveillance, and economic games, 113–15, 194n14
survival of the friendliest, 61–63, 67
sustainability practices, 146, 160–62, 180
Swift, Jonathan, 11, 122
Switzerland: generosity research in, 111–12; Jewish money in banks of, 120–21

tactical deception, 45–46
taxation: evasion of, 64, 96, 155–56; fairness in, 118; free-rider problem in, 144–47; of inheritance, 155–56; policymaking on, 155–56, 158; reducing inequality through, 100, 154–56
Tax Observatory, 117
technology regulation, bottom-up approach to, 138–39
temptation, 102, 105
Tennessee Valley Authority, 139
Tennyson, Alfred, 18
territory, as resource capital, 75
Tether, 127
Tett, Gillian, 175
Thackeray, William Makepeace, vii–viii

Index

Third Wave, 133–35
Thomas, Danny, 125
Three Impostors, The (Machen), 181
tit-for-tat model, 54, 56
Tolkien, J.R.R., 181
Tolstoy, Leo, 64, 104–5, 130
tools: of ancient humans, 15–16, 21–23; of primates, 21
top-down policymaking, 137–43, 158–59
Tower of Babel, 57
TP53 gene, 92
tragedy of the commons, 140–44
Treaty on the Functioning of the European Union, 117
trees, pine, 92
Trivers, Robert, 50, 52, 55
Trolley Problem, 54
true altruism, 90–91, 106–7
Trump, Donald, 81, 137, 165, 182
Truss, Liz, 118
trust, 166–77; blind, 166; credibility in, 167–70; disinformation in absence of, 156; false indicators of, 157, 162, 174; in history of cooperation, 53, 56, 61–62; how to deserve, 174–77; how to place, 167–74, 180; intention in, problem of reading, 66; lack of, effects on policies and politics, 156–62; lie detection and, 60, 167–68; mirroring behaviors in, 42; motivations in, 169–70, 172–74; resource inequality and, 151, 156–57; signaling of, 61–62, 157–59; state surveillance and, 113–14; top-down vs. bottom-up policymaking in, 158–59
Tullock, Gordon, 137–38, 163, 166
Turgenev, Ivan, 12
turtles, 46–47, 84–85

Twa people, 27
Twitter (X), 107–8, 168
two-state solution, 159–60

Ukraine, Russian invasion of, 64
United Kingdom: Covid-19 pandemic relief in, 145; culture of inequality in, 151; economic inequality in, 148–49; fairness in taxation in, 118; Humanists U.K. in, 122, 195n29; inheritance tax in, 155–56
United Nations Institute for Training and Research, 176
United States: climate policy in, 178–79; economic games in, 107–8, 113–14; electricity regulation in, 139–40; fairness in taxation in, 118; inheritance tax in, 155–56; presidential elections in, 125, 137
University of Zurich, 111
Unrest (film), 186n12

vaccine mandates, 158–59
vampire bats, 52
Vanity Fair (Thackeray), vii–viii
Verra, 161
violence, inequality and, 151, 153
virtue signaling, 64–66
viruses, 5
visibility. *See* invisibility
visible rivals, x
Vonnegut, Kurt, 167

war heroes, 80
Washington Post, 106
water resources, 141–42
We (Zamyatin), 164, 167, 181
wealth: of billionaires, 76–78, 89, 149, 155–56; influence on

wealth (*continued*)
  policymaking, 158; relational, 27–29, 79–80; types of, in hunter-gatherer groups, 87. *See also* resource capital
wealth inequality. *See* resource inequality
Weigel, Jonathan, 146–47
WEIRD research, 110, 194n10
welfare programs, free-rider problem in, 145, 178, 179
Wells, Jonathan, 151, 156
West, Kanye, 79
West-Eberhard, Mary Jane, 60
Whately, Richard, 44
white hat bias, 172
Whiten, Andrew, 45–46
"Why Women Aren't Funny" (Hitchens), 124
Wiessner, Polly, 79, 111–12
Wilkinson, Richard, 151
Williams, Bernard, 173
Wilson, Edward O., 2, 35
Wing, The, 125, 126
Wittgenstein, Ludwig, 106, 167
Wodaabe people, 83, 84
women: aggression against, 151; equality for, 123–25; exploitation of, 30–31, 38–39, 123–24; in hunter-gatherer groups, 30–31; in matrilineality, 87, 192n40; sense of humor in, 124; sexism against, 116–17, 123–24
World Economic Forum, 149
World War II, Nazi Germany in, 120–21, 131–33, 137
Wrangham, Richard, 21, 22, 62, 189n29
Writers Guild of America (WGA), 138–39

X (Twitter), 107–8, 168
xenophobia, 95
Xi Jinping, 89

Yamagishi, Toshio, 113–14, 194n10
Yanomami people, 38
yardstick view of fairness, 116–17
Yélî Dnye language, 24
YouGov, 161, 198n49

Zahavi, Amotz, 58–59, 65
Zamyatin, Yevgeny, 164, 167, 181
Zeller, Florian, 104
Zoshchenko, Mikhail, 131, 140, 162
Zuckerberg, Mark, 89